Other Books Available at Holloway.com

The Holloway Guide to Remote Work
Katie Womersley, Juan Pablo Buriticá et al.
A comprehensive guide to building, managing, and adapting to working with distributed teams.

The Holloway Guide to Equity Compensation
Joshua Levy, Joe Wallin et al.
Stock options, RSUs, job offers, and taxes—a detailed reference, explained from the ground up.

The Holloway Guide to Technical Recruiting and Hiring
Osman (Ozzie) Osman et al.
A practical, expert-reviewed guide to growing software engineering teams effectively, written by and for hiring managers, recruiters, interviewers, and candidates.

The Holloway Guide to Raising Venture Capital
Andy Sparks et al.
A current and comprehensive resource for entrepreneurs, with technical detail, practical knowledge, real-world scenarios, and pitfalls to avoid.

Land Your Dream Design Job
Dan Shilov
A guide for product designers, from portfolio to interview to job offer.

Founding Sales: The Early-Stage Go-To-Market Handbook
Pete Kazanjy
This tactical handbook distills early sales first principles, and teaches the skills required for going from being a founder to early salesperson, and eventually becoming an early sales leader.

Angel Investing: Start to Finish
Joe Wallin, Pete Baltaxe
A journey through the perils and rewards of angel investing, from fundamentals to finding deals, financings, and term sheets.

Junior to Senior

Junior to Senior

CAREER ADVICE FOR THE AMBITIOUS PROGRAMMER

David Glassanos

A practical guide to self-confidence, personal growth, teamwork, learning, communication, and delivering results—the soft skills that every programmer needs to thrive in their job and be ready for a senior role.

WES COWLEY, EDITOR

HOLLOWAY

Published in the United States by Holloway, San Francisco
Holloway.com

Cover design by Order (New York) and Andy Sparks
Interior design by Joshua Levy and Jennifer Durrant
Print engineering by Titus Wormer

Typefaces: Tiempos Text and National 2
by Kris Sowersby of Klim Type Foundry

Print version 1.0 · Digital version e1.0.1
doc 6060ff · pipeline f2e7e1 · genbook 34a37b · 2023-10-25

Want More Out of This Book?

Holloway publishes books online. As a reader of this special full-access print edition, you are granted personal access to the paid digital edition, which you can read and share on the web, and offers commentary, updates, and corrections. A Holloway account also gives access to search, definitions of key terms, bookmarks, highlights, and other features. Claim your account by visiting: **holloway.com/print20398**

If you wish to recommend the book to others, suggest they visit **holloway.com/jts** to learn more and purchase their own digital or print copy.

The authors welcome your feedback! Please consider adding comments or suggestions to the book online so others can benefit. Or say hello@holloway.com. Thank you for reading.

The Holloway team

LEGEND

Some elements in the text are marked for special significance:

◇ IMPORTANT	Important or often overlooked tip
◈ CAUTION	Caution, limitation, or problem
⚑ CONTROVERSY	Controversial topic where informed opinion varies significantly
☙ CONFUSION	Common confusion or misunderstanding, such as confusing terminology
⟫ EXAMPLE	An example or illustration
⚯ RESOURCES	Additional readings or resources

Web links appear as numbered footnotes in print.

References to other related sections are indicated by superscript section numbers, prefixed with §.

TABLE OF CONTENTS

1 Introduction

This book is about accelerating your career as a programmer. It is not about programming.

Moving up the ladder from a first or second job to a senior role requires a set of skills that coding boot camps and computer science degrees don't typically cover and that programmers often have to teach themselves.

There are plenty of books that offer new ideas and technical concepts to help you write better code. You can find books and courses that cover the ins and outs of a programming language, and tutorials that show how to build something while learning a new language or framework.

But much of the work you do during your career will be planning, reviewing, strategizing, collaborating, and many other things that don't involve coding. To excel in your career as a programmer requires *soft skills* that are different from your technical abilities. As you'll learn throughout this book and throughout your career, the soft skills are often what set you apart from other programmers and increase the impact of your work. They will factor into your manager's decisions when it comes time for a promotion. And the best part is that almost none of them will be outdated in a few years, even as the industry evolves.

The *practice* of programming may feel like an individual activity, but the *process* of building software involves collaborating with other people, both technical and nontechnical. While reading, writing, and debugging code will challenge you, possibly the hardest thing you learn over the course of your career is how to work effectively with other people. You learn how to compromise and collaborate with people that may have different personalities and opinions than you do. You learn how to communicate your ideas and listen to other people's points of view. All these skills are needed in a senior role.

Unfortunately, there aren't many books that teach these soft skills. Many of these skills are developed with experience. You'll make mistakes along the way, and that's okay, because everyone does. What matters most is that you learn from your failures and gain the wisdom to avoid those same mistakes in the future. And you may learn enough to pass that wisdom along to others so they can learn to avoid similar situations. In fact, that is one that sets a senior programmer apart from others—the ability to lift their team to greater heights.

1.1 *Why I Wrote This Book*

I started my programming career at a small startup called Mertado, soon after they completed the Winter 2010 batch at Y Combinator. Nine months into my new role, we were acquired by Groupon to help build out their Groupon Goods platform soon after they had gone public. After the dust had settled, I was placed on the team that built the machine learning system to personalize 100 million emails sent to inboxes each morning. I was responsible for taking the personalized outputs from the MapReduce jobs, rendering the results in our HTML email templates, and performing multivariate tests on different audiences in order to increase the conversion rate, where a one percent increase meant millions of dollars in additional revenue.

It turns out that working at a small, ten-person startup is vastly different from working at a public company on a large engineering team. Not only was I exposed to new technologies, development workflows, build systems, and enormous codebases, I also got first-hand experience observing how high-performing teams deliver software at scale. I started to see how all of my coworkers juggled different tasks and priorities, and as I observed, I started to learn from them.

What I noticed early on was that they didn't just sit there and code all day with their headphones on. Yes, my coworkers delivered clean and robust code at a fast pace—but they were good at the entire software development process. The most impactful engineers combine technical depth with a broad set of soft skills, people skills, and product skills.

I've made many mistakes throughout the course of my career. It took hard work and trial and error to learn how to navigate office politics, manage risk, and work well with others to deliver quality software. I had to learn most of these soft skills on the job and by observing others as I navigated my career.

But I wish I'd had better resources to prepare me for the obstacles on the path to becoming a senior software engineer. My goal in writing this book is to pass along the knowledge I've gained so far in the hope that it will help the next generation of programmers be team players and build fulfilling careers.

1.2 *Who This Book Is For*

This guide is focused on providing junior and mid-level programmers the tools they need to excel in their careers—the journey from first full-time engineering job to earning a first promotion.

This book is primarily for individual contributors: it does not cover the broader topic of managing other junior and mid-level programmers—but engineering managers and senior engineers may still find some of the material useful for mentoring or sharing with their team members or direct reports.

- **New programmers.** If you're just starting to learn how to program by teaching yourself, or you're enrolled in school or a coding boot-camp, congrats! You're just getting started on a fun and exciting journey. While this book won't teach you how to program, it will prepare you for what to expect when you land your first full-time programming job. There's a lot to learn, so be patient and take things one step at a time. You got this!
- **Junior programmers.** If you've already landed your first full-time programming role, nice job! You've proven that you can build programs and solve difficult problems, so now it's time to focus on being a team player. You may already know some of the topics covered in this book, but it never hurts to reinforce those ideas and build good habits. You'll be reading a lot of unfamiliar code and asking questions, so you may find those sections especially useful.
- **Mid-level programmers.** If you've been working as a full-time programmer for a few years, you're probably starting to get the hang of things. You may know your way around the codebase and be comfortable working on a team, so some of the topics covered in this book may already be second nature to you. This book will still be helpful if you use it to focus on honing those skills even further to solidify those good habits. The latter half of the book should be especially useful to mid-level programmers as you begin to make more impactful technical decisions.

- **Senior programmers, mentors, and managers.** We all had to start somewhere in our careers, and by now, almost all of the information in this book should be familiar to senior programmers and programming managers. It's easy to forget just how far you've come and how difficult it was to learn certain skills. This book offers perspective to experienced programmers and reminds them of areas they can offer assistance in when mentoring and managing junior and mid-level programmers.

1.3 *What Is Covered*

This book is meant to be a resource for junior and mid-level programmers on soft skills required to excel as a professional programmer. It's meant to be your career guide as you learn to navigate the workplace, from learning the lingo and how the business operates to working as part of a team producing quality software and real value for your customers.

The book is not necessarily designed to be read from cover to cover, so feel free to skip around to different sections if a topic seems particularly relevant to you at the moment.

My goal is to help you establish a personal roadmap to guide you as you start your programming career or reach that next level of responsibility. I hope this becomes a book you revisit regularly to review your progress and recalibrate your priorities for your next career goal.

Let's look at what we'll cover:

- **Growing Your Career.**[§2] In the first section, you'll start off by learning how and why you should set expectations for your career. You're at the very beginning of a years-long journey to becoming a senior programmer, so it's important to set some long-term goals to work towards.
- **What Makes You a Senior Engineer?**[§3] In this section, we'll dive into specific characteristics that differentiate a senior engineer from a junior engineer. You'll gain a better understanding of what you should focus on if you're working towards a promotion to a senior title.
- **You're Not an Impostor.**[§4] Next, you'll learn that everyone deals with impostor feelings during their career. You'll learn how to recognize and understand these feelings as well as ways to reduce them when you're overwhelmed.

- **Working with Your Manager.**[§5] You don't have to be friends with your manager. But, they'll have a big influence on your career, so it's important that the two of you work well together. In this section, you'll learn how to understand your manager's style and align your goals so both of you can succeed.

- **How to Recover from Mistakes.**[§6] Even the best programmers make mistakes. What sets the best apart is how they respond when things go wrong. In this section, you'll learn what you can do when you make a mistake and how you can recover quickly.

- **How to Ask Better Questions.**[§7] Asking good questions is a simple and effective way to supercharge your learning early in your career. Here, I share tips and techniques for learning as much as you can from your coworkers and for tapping into their experience.

- **How to Read Unfamiliar Code.**[§8] You'll read lots of other people's code throughout your career, which will have a big influence on how you write your own code. In this section, you'll learn how to get up to speed quickly in an unfamiliar codebase.

- **How to Add Value.**[§9] One of your main responsibilities as a programmer is to add value for your customers and for your employer. In this section, we'll cover specific ways in which you can do this.

- **How to Manage Risk.**[§10] Another top priority for programmers is to identify areas of risk and manage them in a way that's consistent with business objectives. We'll cover types of risk common in software development and ways you can help mitigate them in your own role.

- **How to Deliver Better Results.**[§11] At the end of the day, your job is to ship code and deliver value to your customers. This section offers specific things you can do to work efficiently and to increase your productivity when writing code.

- **How to Communicate More Effectively.**[§12] Communicating well sounds simple but can be deceivingly difficult in fast-paced work environments. In this section, you'll learn how to improve your speaking and listening skills to better convey your ideas and stand out among your peers.

- **Work-Life Balance.**[§13] Life isn't all about work, so in this section, we'll go over the importance of creating a work-life balance that works for you. We'll cover ways in which you can expand your horizons and prevent burnout that can affect your productivity and the quality of your work.

- **Asking for the Promotion.**[14] This is it. This is where you ask the big question to your manager. You'll learn what you should do to prepare for this moment and how you should approach the conversation. At this point, you'll be ready to take on more responsibility.

By the end of this book, you should have a deeper understanding of the soft skills required to succeed as a software engineer, along with a path forward for applying those skills as you work towards a promotion to a senior role. At that point, it's up to you to put these ideas into practice and make it happen. So, with that said, let's get started!

1.4 *Getting More Out of This Guide*

1.4.1 LEGEND

Key points are highlighted like this:

◇ IMPORTANT   An important note.

◈ CAUTION   A caution.

⚸ CONFUSION   A confusion or reminder.

In addition, you'll find examples highlighted:

❯ EXAMPLE

An example or scenario.

1.4.2 COMMENTS AND IMPROVEMENTS

◇ IMPORTANT   If you're reading this Holloway Edition of the book online, please remember you can add comments and suggestions. This will help the guide improve in future revisions, and your helpful comments will be published to assist other readers.

1.5 *Acknowledgments*

This book is only possible because of the mountain of support from family, friends, coworkers, and new acquaintances I've met through people involved with this book.

I want to thank my parents for their unwavering support, for always believing in me, and for providing the foundation for me to build a life and a career.

Thank you to my brother, Brian, for inspiring me to begin writing down my thoughts on how to help people build a successful programming career.

Thank you to all my current and former coworkers who have had an impact on me, big or small, throughout my programming career.

To my current and former managers who have taught me so much about all aspects of planning, designing, building, and maintaining software systems. The process can be messy at times, but you've taught me how to embrace the chaos and trust the process.

To Mehul Shah, for taking a chance on me and kick-starting my professional programming career.

A special thank you to everyone who helped review and revise early drafts of this book. It's hard to put into words how much better this book turned out thanks to the following people: Josh Levy, Wes Cowley, Rachel Jepsen, Robert Swisher, Tim Sorweid, Yuri Subach, Ozzie Osman, Nitin Sharma, Ryan Rusnak, and anyone else I've regrettably forgotten.

And finally I want to thank my wife, Julia, for all her love and support, and for cheering me on every step of the way.

2 Growing Your Career

You're here because you want to learn how to be a better software engineer. You may already know how to code, or you might still be learning your first programming language. But that's not what this book is about. This book won't teach you how to write cleaner code or architect better systems. While the ability to write elegant and scalable programs are important skills to have as a programmer, they alone do not make you a successful programmer.

This book will teach you everything *else* you need to be a well-rounded programmer—a team player who collaborates well with others, who communicates effectively, and who knows how to manage risk while delivering value to customers. These skills, called soft skills, become increasingly important as you climb the career ladder.

In fact, while many programmers reach a certain level of technical ability during their career, not all of them possess the soft skills necessary to have the greatest impact on a team. Simply put, these soft skills are what separate good programmers from great ones in the later stages of their careers. They will help you unlock new opportunities as you take on increased responsibility and accountability.

Unfortunately, many programmers neglect these soft skills during their early career because they're laser focused on improving their technical skills. As a result, they often have to change their habits as they learn how to collaborate, communicate, balance trade-offs, and lead projects—and changing old habits can be hard.

The good news is that these soft skills can be learned. If you put in the effort to build good habits now, over time they'll become second nature. Some of these skills may take years to develop, so it's important to start right now, and persistence is key. If you're willing to put in the work to build good habits now, it'll pay dividends for the rest of your career.

As a programmer, it's important to focus on your career goals because if you don't know where you're going, you won't know when you've reached your destination. Even the most ambitious programmers won't get very far if they don't have an end goal in mind. Setting a North Star will keep you focused, motivated and on track as you navigate the twists and turns of your career.

◇ IMPORTANT   It's entirely okay if you're not sure what your career goals are at this time. It can be difficult to know what you want from your career while you're just getting started—this is common. This exercise is meant to help you think about your future goals, but don't feel like you have to decide right now where you want to be in 10 years.

There are a few different aspects you should consider when thinking about your career goals—the engineering path you want to take, whether you want to focus on a generalized set of skills or a specialized one, and which leadership path you want to pursue. Together, these factors help you determine where you should focus your time and energy during your career development.

Let's take a look at each of these in more detail.

2.1 *Your Engineering Path*

You've probably already put some thought into the engineering path you want to take. Perhaps you're learning your first programming language, or you've found a framework you like and have built a few projects with it already. That's great, because that's the first step to choosing your engineering path.

The engineering path you follow will determine your journey to increasing technical mastery of one or more technologies of your choosing. The technologies you decide to learn early on in your career may seem inconsequential right now, but they will have a drastic impact on your career.

In some cases, the technologies you chose to focus on will dictate which companies and industries you'll apply to when searching for a job. On the other hand, if you've already landed a job, you'll want to do as much as possible to become an expert in the technologies your company has invested in.

Let's dig a little deeper into each of those scenarios.

2.1.1 CHOOSING TECHNOLOGIES TO SPECIALIZE IN

Programmers have a natural tendency to seek out roles that require skills similar to what they already possess. Part of this is because interviewing is hard, so people tend to want to interview in a language or technology they are comfortable and confident with. Because of this, candidates will research and apply to companies that work with technologies similar to ones they are familiar with. So, while it may not be obvious, the technologies you choose to learn early in your career will determine which companies and industries are within reach when it comes to searching for a new role.

🏳 CONTROVERSY Whether companies should filter candidates based on their experience in a specific language, like JavaScript or Rust, is a controversial subject. Good programmers can, of course, learn new languages. But the practical reality is that many businesses have limited time and resources when it comes to hiring and onboarding new employees. They can't afford to spend more than a few weeks or months getting a new hire up to speed, so they tend to recruit programmers that are experienced in specific technologies the company has invested in.

If a company's product is built on a codebase consisting of Python code, for example, they're more likely to consider candidates with strong Python skills over people with no Python experience at all. Taking the example further, if the company utilizes a specific framework or library such as Django or TensorFlow, they will naturally seek out programmers who are familiar with these technologies. The more a new hire is required to learn during onboarding, the longer it will be until they can start to provide real value, so it's more cost-effective for companies to hire candidates who already possess the team's desired skills, rather than hiring someone and teaching them those skills.

Most companies tend to be more lenient when it comes to frameworks versus languages, because frameworks are easier to learn if a new hire is already familiar with the underlying programming language. At that point, it becomes a matter of learning the different APIs of the framework, which is arguably easier than learning the semantics of a new programming language in addition to those APIs.

Which brings us back to what we talked about earlier—that it's easier to land interviews and get offers at companies seeking experience with technologies you're familiar with. While it's still possible to find companies willing to hire candidates with no prior experience in the needed programming language, those companies can be difficult to find.

However, as a junior programmer, you are presented with a unique opportunity, because there are companies that are willing to hire candidates in the early stages of their career even if they don't have any prior experience with specific technologies. These companies tend to be established enterprises with the budget and resources to invest in their employees over a longer timeline. They tend to operate in niche industries or ones with high barriers to entry, so they can afford to spend a year onboarding a new programmer. Some industries, such as aerospace, healthcare, energy, and many others, may take years to get up to speed in, so those companies are willing to invest in their new hires and teach them everything they need to know.

Startups and small businesses, on the other hand, don't have the luxury of large budgets and longer timelines that established corporations can afford. Startups naturally need to move quickly as they seek to validate and scale their business model, so they tend to focus their hiring on candidates that have the specific skills they need. Smaller companies need new hires to add value as fast as possible, so it may be harder to land an interview at a startup if you don't have experience with a specific technology.

So, how do you go about choosing which technologies to specialize in? There's no easy answer to that question because everyone's situation is unique, but I'll try to give some advice that will help you decide.

First, think about the languages and frameworks you're already familiar with. Are you comfortable enough that you feel you could interview for a role in that language? Try solving some practice problems on Hacker-Rank[1] or LeetCode[2] to gauge how ready you are. If you feel like you're prepared enough to solve interview questions on the spot, then you're ready to start applying for roles.

Start searching on LinkedIn[3] or other job boards for companies that are looking for programmers with experience in your language. You may even be able to get your foot in the door for interviews with startups and smaller companies that need to hire programmers with your specific skill set.

If you don't think you're ready to interview with a specific language yet, that's okay. That just means you'll need to keep practicing and learn the language until you feel you're comfortable enough to interview. You're not completely out of luck though, because you can still look for larger enterprise companies that are willing to hire smart candidates without any prior experience with a language or technology. These companies tend to hire based on your potential rather than how well you know a specific skill or algorithm.

Now, let's look at what you should consider once you've landed a role and are invested in a specific technology.

---

1. https://www.hackerrank.com/
2. https://leetcode.com/
3. https://www.linkedin.com/

2.1.2 MASTERING A TECHNOLOGY

Getting hired for your first programming job is one of the biggest hurdles of your career. The next step is to focus on mastering the languages your team uses as best you can.

Unfortunately, you may not have much of a choice when it comes to choosing which programming languages and technologies you will work with. The decision was probably made for you by other programmers who built the company's systems before you joined. Even if you're familiar with the programming language they chose, it's possible they built the system using a framework or library that you've never used before.

Like it or not, your ability to succeed is dependent on how well you know the technologies that your team uses. As you'll learn in the next section,[§3] your technical mastery of these tools can actually have an effect on whether or not you get promoted to a role with increased responsibility.

As you gain more experience and work on larger projects, you'll need to understand the technologies you work with at a deeper level in order to design and scale solutions to meet the business' needs. You can't pick the right tool for the job if you don't have a good understanding of the strengths and weaknesses of each technology you're working with.

So, what are some ways in which you can master the technologies you work with?

The most obvious resource that you can leverage is right in front of you—your coworkers. They are likely the ones who have written large portions of the codebase you're working in, which means you can lean on them to learn the best practices and core concepts.

Most programmers need some help early in their career as they make the transition from learning how to program to learning how to solve complex problems through code. It's a small but important difference that requires a different way of thinking about problems. You've learned how to use the foundational concepts of your language to build programs, but now you need to learn how to break down a problem and construct a solution using the building blocks available to you.

This transition can be difficult if you try to do it entirely on your own, so it's best to try and leverage the experience and knowledge of your coworkers as much as you can early in your career. Read their code, review their pull requests, and ask questions about how they come up with their solutions. There's a lot that you can learn from reading code[§8] and asking the right questions,[§7] so we'll cover those topics in depth in later sections.

With the help of your coworkers, you'll be on a good path to mastering your first programming language. Early in your career, it's important to stay focused and get really comfortable with the language you're using. Don't worry too much about learning other languages at this stage in your career. Your priority should be to get really good at one language, because you'll always be able to fall back on that as your foundation. Assuming you chose a language that has a large community and is used by many companies, you'll have job security because you'll always be able to find work at companies looking for your skill set.

It usually takes a few years to reach an advanced level with a programming language. Once you approach this point, you have to start thinking about where you want to go in your career. One decision you'll have to make is whether to take a generalized approach with your skill set or a specialized one.

Let's take a look at both of these options.

2.2 *Generalist vs. Specialist*

At some point, you'll need to decide whether you want to branch out and learn additional languages and technologies, or double down and become an expert in one or two specific ones. Programmers can be successful in either path, so it mainly comes down to personal preference and some careful consideration about where you want to take your career.

Let's look at each path in detail.

2.2.1 GOING BROAD

A programmer who views themself as a generalist typically wears many hats and will have an intermediate-to-advanced knowledge of many different complementary skill sets. They may work with many different programming languages, databases, and platforms. Their range of skills allows them to collaborate across multiple disciplines to solve customer problems.

🐾 CONFUSION It's worth noting that the term "generalist" does not mean "full stack," even though the two terms are often conflated. While full-stack developers are considered generalists, you can still be a generalist within a vertical such as mobile development, game development, or systems programming.

Generalists typically cast a wide net, which gives them the ability to jump around to different projects, companies, and even industries throughout their career. They bring with them a broad set of experience because they've worked on many different projects and technology stacks. In some cases, they can bring experience from one project or industry over to another in order to solve problems in novel ways.

Because generalists carry such a diverse set of skills with them, they have the advantage of a larger job market and have an easier time finding work. Many companies prefer to look for candidates with a generalist skill set because they deal with constantly evolving requirements from customers. New projects arise from changing requirements, and companies are able to shift generalist developers from project to project without having to hire new programmers with a specific skill set.

More often than not, it's a good idea to generalize your skill set during your career, but only after you've become comfortable in your first programming language. When you generalize, you are exposed to many different technologies and industries, but just because you've taken the generalist path does not mean you can't specialize in one area. In fact, starting off as a generalist gives you the opportunity to try a number of different technologies before deciding which one you'd like to specialize in.

◇ CAUTION   One thing generalists should consider is the need to stay up-to-date on multiple skill sets as the ecosystem for each technology evolves in parallel. Generalists may fall behind in one or more technologies if they don't find time to practice those skills.

Now, let's look at the specialist option in more detail.

2.2.2 GOING DEEP

The specialist path is an option for those programmers who want to take a deep dive on a technology, effectively becoming an expert in one specific area. You may choose to specialize in a programming language, a database technology, or a field of computer science such as machine learning or algorithms. It can take years, sometimes decades, to reach an expert level, but once you do, you're often rewarded with job security and higher salaries, because companies will compete for your skills if they're in high demand.

> ❯ EXAMPLE

Examples of specialists include a computer vision expert working for an autonomous driving car company, someone who specializes in quantitative trading systems working at a hedge fund, a big data expert working for an IoT company, or even an expert in the Ruby programming language working for a company that has built their codebases on Ruby.

Specialists often have leverage when it comes to negotiating job offers and salary increases because their deep knowledge in a field may be considered a competitive advantage in certain industries.

Although specialists may not have trouble finding new work, the size of their overall job market may be much smaller than that of a generalist because there are fewer companies looking for their specific set of skills. This presents an interesting problem and something that should be considered when thinking about your career path: what if you specialize in the wrong thing?

There's inherently more risk if you decide to specialize in a programming language or field of computer science because you'll need to devote years of your life becoming an expert in one specific set of skills. If the industry evolves during that time, you may be putting energy into something that isn't guaranteed to be in demand a few years from now.

That's a calculated risk you'll need to take when deciding what to specialize in, but if you understand a specific technology well enough to become an expert in the first place, you may be able to anticipate where the industry is headed and get a head start. When done right, specializing can be very lucrative for programmers.

⚙ CONFUSION While most programmers think about specializing in a specific technology, it's also possible to specialize in a specific industry. You may use different technologies as you jump from company to company within the same industry, such as healthcare or finance. Although you may not be an expert in the technology at a company, you bring a wealth of industry knowledge with you, which can be just as valuable. When it comes to hiring, a company may consider years of experience in an industry as important as years of experience with a technology.

Now that we've gone into detail about the technical path of your career, let's look at another aspect: leadership.

- Generalists vs specialists - who has a greater chance of success?[4] (abhijitbhaduri.com)

2.3 *Your Leadership Path*

In the early stages of your career, you're probably going to be focused on developing your technical skills first before thinking about any kind of leadership path. While it's good to have a sense of where your career is heading from a technical perspective, it's equally important to start thinking about your career from a leadership perspective.

---

4. https://www.abhijitbhaduri.com/2019/08/15/generalists-specialists/

At some point, usually after you've been in a senior engineering role for a while, you'll reach an inflection point where you'll need to decide if you want to stick with the technical track as an individual contributor (IC) or make the jump to the management track. There are excellent leadership opportunities in both paths, but it's important to understand what those responsibilities look like as you think about the direction you want to take your career.

Here's what the paths look like at a typical tech company.

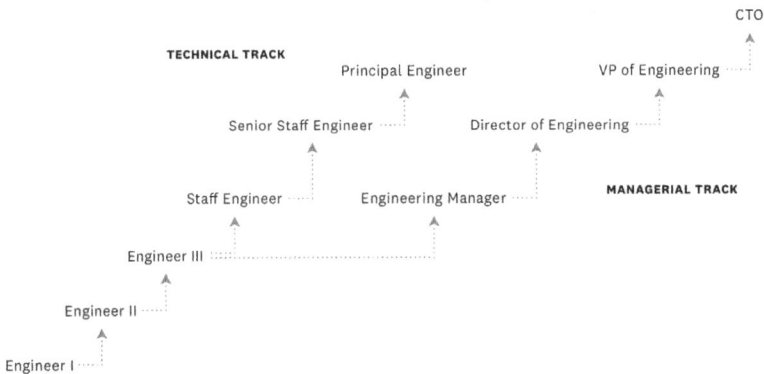

Source: Holloway Guide to Technical Recruiting and Hiring.[5]

All programmers begin their career on the technical track as individual contributors, because the foundational skills you develop early on are prerequisites for both the technical and management tracks.

The primary responsibility of an individual contributor is to support the organization through the projects and tasks you work on. You won't carry any management responsibilities or have anyone report directly to you, so you will only be expected to manage yourself and your own work.

Most programmers enjoy working as an individual contributor because it allows them to do what they love—programming. The majority of your time will be spent reading, reviewing, and writing code. You'll design and implement feature enhancements and identify and fix bugs. As you start out, you will rely on senior engineers to provide direction and context for the changes you make, but as you gain more experience and autonomy, you'll shift to designing and implementing your own solutions to well-understood medium- and large-sized problems.

5. https://www.holloway.com/s/trh-job-titles-levels-fundamentals-for-software-engineering

Once you reach this inflection point, you'll need to decide if you want to continue down the technical track or shift to the management track.

Let's look at both tracks in more detail.

2.3.1 TECHNICAL TRACK

After you've reached a senior engineering role, the most common levels on the technical track are the Staff Engineer, Senior Staff Engineer, and Principal Engineer roles. These deal with increasing job complexity and often require expert knowledge in multiple technologies in addition to development best practices.

In contrast to the junior, mid-level, and senior engineers, the higher roles on the technical track are responsible for exploring different solutions to large and poorly understood problems. They often build proofs of concept to demonstrate the feasibility of different solutions before choosing which direction to pursue.

Staff and principal engineers are expected to perform expert programming tasks and find opportunities for large-scale refactorings in order to reduce technical debt. They think strategically about how the technical systems fit into the rest of the company in order to provide leverage and opportunities for the company to scale.

They may not necessarily sit on just one team. Staff and principal engineers often move across teams to wherever difficult and vague problems exist. They may design a solution for one team and then move to a different team to help them find the best direction for a different problem.

Although theirs are primarily technical roles, most senior programmers on the technical track provide leadership through their technical guidance on projects and through mentoring a handful of individuals. Many engineers continue as individual contributors on the technical track because they incorrectly assume that they won't be managing people. While it's true that staff engineers don't have direct reports, there is a fair amount of people management involved. They still need to work with other engineers to build consensus and buy-in for the ideas, concepts, and initiatives they believe the company should pursue.

The technical path provides opportunities for those who aren't necessarily interested in managing people, while still allowing them to practice their technical skills in order to design and build systems that provide value for the organization.

2.3.2 MANAGERIAL TRACK

The path toward management is gradual. You don't suddenly become a manager one day when the title gets assigned to you. Rather, it takes years of experience leading projects from a technical point of view, learning to collaborate with others, and listening to the needs of other programmers. Engineering managers have a knack for helping other programmers achieve their potential, and that isn't something that can be learned overnight.

> ❯ EXAMPLE
>
> Common roles you'll find in the managerial track consist of Lead Engineers, Engineering Managers, Directors of Engineering, VP of Engineering up to the CTO or CIO of the company.

Rather than continuing down the technical path, some programmers may instead find themselves more comfortable leading the direction of projects or the people involved in them. While it's possible they still contribute code here and there, they enjoy fostering the collaboration and communication between engineers working on a project to get it across the finish line.

While managers are expected to have an intimate understanding of the technologies involved in the company's products, they are also required to have an expert knowledge of the products themselves, including how those products are implemented from a technical perspective.

In addition to balancing tactical short-term goals with strategic long-term goals for the company, to be a great leader in the managerial track you need to be able to recruit and retain great engineers. Part of the manager's job is to build great teams, and in doing so, they need to find ways to foster career growth for the engineers on their team. Above all else, managers support the needs of everyone on their team so that they can all achieve their full potential.

◇ IMPORTANT While the leadership path consists of two discrete tracks (technical and managerial), engineering managers almost always start out as individual contributors. Just like how managers need to understand the technical issues at hand, individual contributors need to understand the management issues in order to be effective.

- The Secret to Growing Your Engineering Career If You Don't Want to Manage[6] (effectiveengineer.com)
- Why All Engineers Must Understand Management: The View from Both Ladders[7] (hackernoon.com)

2.4 *It's a Marathon, Not a Sprint*

Now that we've looked at what things to consider when thinking about your engineering path—whether to generalize or specialize, and options for your leadership path—it may feel like there are a lot of things to think about. There's a lot that goes into software development, but you don't need to decide all of these things at once.

It may be overwhelming to think about all of this right now, so it's important to give yourself a reasonable timeline when it comes to advancing in your career, and to keep things in perspective when setting goals for yourself, because you have a long career ahead of you.

Some skills take years to learn and decades to master, so be patient and take things day by day. Continuous improvement is the best thing you can focus on right now, as that will give you a strong foundation you can build on. Small improvements every day will compound over time, and soon you'll look back and be surprised at how quickly you're progressing.

Think about your career as if you're running a marathon and you're just starting your race. You wouldn't want to sprint to the finish line right now because you might burn yourself out. Just be patient, and take it one career milestone at a time.

2.5 *Go at Your Own Pace*

Additionally, try to focus on running your own race. At this point in your career, you're competing only against yourself, not others, so try your best not to put pressure on yourself if your peers are progressing in their

---

6. https://www.effectiveengineer.com/blog/secret-to-growing-software-engineering-career

7. https://hackernoon.com/
why-all-engineers-must-understand-management-the-view-from-both-ladders-cc749ae149
05

careers at different paces than you are. Even though it may seem like you're running the same race, each person begins at a different starting line and is running to their own finish line. It won't do you any good to compare yourself to others because you're not even running the same race.

◇ CAUTION Keep in mind that as you get promoted and move up the org chart, you will have to compete against others as the number of available roles begins to narrow. You'll have to shift your mindset from competing against yourself to competing against others, and this is where the soft skills become even more important.

People learn at different rates, and what may come easy to one person might take weeks or months for someone else to grasp. Learning a new skill often requires you to change your way of thinking, sometimes forcing you to change how you approach a problem. While it might click right away for some people, try not to get discouraged if it takes you a little longer to learn a new technology.

You should be proud of every raise, promotion, job offer, or other career milestone regardless of how long it took you to reach it. These things take a lot of hard work—and more importantly, a lot of patience. Just focus on growing each day as a programmer and as a person, because you're in control of your career and only you get to decide when you've crossed the finish line.

𝒪 RESOURCES

- You Don't Have to Manage, But You Still Have to Lead[8] (somehowmanage.com)
- The Myth that Technical Skills Alone Will Make You Great[9] (effectiveengineer.com)

8. https://somehowmanage.com/2021/10/22/
   not-being-a-manager-doesnt-exempt-you-from-engineering-leadership/
9. https://www.effectiveengineer.com/blog/interpersonal-soft-skills-in-engineering

3 What Makes You a Senior Engineer?

It's a question that's been debated for decades within the software community—what qualities separate a junior engineer from a senior engineer? In this section, we'll take a closer look at this question and help you understand why the answers vary so greatly. We'll also explore a few topics that are absolutely critical if you're working towards a promotion to a senior role. By the end of this section, you should have a concrete idea of the differences between a junior and senior engineer. Let's get started.

3.1 *Different Meanings*

If you were to pull up your preferred search engine, enter the query "what is a senior software engineer?", and hit enter, you'd be presented with hundreds of different explanations of what it means to be a senior software engineer.

It's interesting that as programmers, we work in an industry that has self-organized around standards and protocols such as TCP/IP, HTTP, HTML, IMAP, SMTP, and many others that play an important role in a connected society. These standards provide a basis for which we all agree on something—a mutual understanding. And yet the truth is that there are so many different interpretations on how to define a senior engineer precisely because there is no standard definition for what is actually required in order to call yourself one.

The software industry is still very young when you look at it in the context of history. While other industries have formed consensus on what standard requirements should be met in order to do business or practice an occupation, we're still learning the best way to deliver software reliably and efficiently. You don't have to pass a licensing exam in order to write software professionally, you just need to demonstrate that you know enough about programming to pass an interview.

The reason for this ambiguity is partly because of the diverse opportunities within our industry. As programmers, we can work for small seed-stage startups all the way up to Fortune 500 public companies, and across almost every industry—energy, healthcare, finance, entertainment, and education, to name a few. So, when you really think about it, how *should* we standardize a common definition? A senior engineer's role and responsibilities at a seed-stage SaaS startup are vastly different from those at a public healthcare corporation, hence we see different interpretations for what it actually means to be senior.

Each business needs to form their own organizational structure around what works best for them, so it makes sense that roles and responsibilities are unique to each enterprise. Yet people try to compare apples to oranges all the time when it comes to debating the characteristics that make up a senior engineering role. In fact, what some people fail to realize is that the word "senior" has multiple meanings here, and each organization interprets the title differently.

3.1.1 YEARS OF EXPERIENCE

Some companies interpret seniority as a reflection of a software engineer's actual years of experience at a company or in an industry. Certain industries are complex enough that it takes years to learn the nuances of a company's product suite and how it solves a customer's pain points. A software engineer who's spent years building a complex product carries vast amounts of domain knowledge with them, and they understand the product and the industry better than anyone else.

While these engineers typically have strong technical skills, they may not always be the most technically skilled on their team, yet they still get promoted. They may bring domain knowledge and experience delivering software in the industry, which some companies value more than pure technical ability.

◇ IMPORTANT You tend to see years of experience valued in highly regulated industries where domain expertise is more valuable than pure technical abilities, such as aerospace, telecommunications, transportation, healthcare, energy, and certain financial businesses.

3.1.2 TECHNICAL SENIORITY

On the other hand, some companies interpret seniority as a reflection of a software engineer's technical abilities. While not always the case, businesses in a growth phase tend to hire or promote from within based on one's technical abilities. This is because these companies have validated their business model and are evolving from a startup into an enterprise. They need to scale their codebase, team, and software development processes as they cross the chasm from an early adopter product to a mass-market solution.

Growth companies need strong technical people to lead the charge as they refactor, rewrite, and scale their systems to handle more demand. Domain knowledge is valuable in these cases, but some of the challenges to be solved are purely technical, so companies may hire and promote from within the candidates with the strongest technical abilities, even if those people have less years of experience than others. This is where it pays to specialize in specific technologies, because you may be able to find companies in a growth phase that are looking to hire for strong technical skills in one or two specific technologies, regardless of whether you have industry experience. For example, a fast-growing security company may need to hire engineers with deep knowledge in cryptography, or a regional internet service provider that's looking to expand into a new market may be looking for network engineering experts.

◇ IMPORTANT You tend to see this more in startups and fast-growing technology companies. If a business is focused on scaling at all costs, pure technical ability is more valuable than domain expertise.

Most companies combine both interpretations and look for well-rounded software engineers that have both years of experience and strong technical abilities. They view seniority on a spectrum, with years of experience and technical abilities at either extreme. Their ideal candidates would bring both domain expertise and technical skills, but they may make exceptions if they come across a candidate that skews towards one or the other.

Now that we've uncovered the different interpretations of what it means to be senior, let's look into some skills required by all senior programmers, regardless of whether the title is based on years of experience or technical knowledge.

3.2 *Dealing with Ambiguity*

When you're first starting out in your programming career, you'll be fixing bugs and extending existing features to add new functionality. Most of what you'll be working on will be determined by a product manager, your engineering manager, or a senior engineer. It'll be up to them to determine most of the functional and technical requirements, and it'll be your job to implement a working solution.

As you grow into a senior role, you'll be faced with more ambiguity in the problems that you need to solve. There won't always be a straightforward answer, and oftentimes there will be a number of ways in which you could solve a problem. It'll be up to you to weigh the trade-offs and determine the best path forward, which may not always be the ideal technical solution.

Senior engineers are often asked to solve difficult technical problems when there isn't always a good understanding about how to get there, such as keeping a user's shopping cart updated in real time across both web and mobile devices. It's up to the senior engineer to break the problem down into smaller, more manageable pieces, determine dependencies between the pieces, and put together a plan for building a solution. Junior engineers, on the other hand, often have a known path set out in front of them and are tasked with working towards a goal, such as making sure orders are processed successfully and inventory numbers are updated accordingly.

Dealing with ambiguity is what sets a senior engineer apart from a junior engineer. This partly comes down to experience, but it also has to do with problem-solving skills, creativity, collaboration, and good communication. It takes all of these skills to keep projects moving forward and finishing on time, and senior engineers learn how to leverage them in order to reduce ambiguity.

3.3 *Dealing with Accountability*

In addition to being able to cut through ambiguity to deliver results, senior engineers are expected to be accountable to others and themselves. They are responsible for project timelines and ensuring that the features shipped to production meet all of the project's requirements.

Junior engineers sometimes blame the quality assurance (QA) engineers or other devs who reviewed their pull requests for not finding a bug in their code, rather than accepting responsibility for it being there in the first place. Senior engineers, on the other hand, accept that responsibility and take full blame for letting bugs slip through, even the ones that aren't caught by peer reviews and QA engineers.

When senior engineers make mistakes, they treat their peers as teammates, rather than adversaries. Being able to accept responsibility for your mistakes and work with your team to identify and fix the root cause is a sign of maturity, and it shows that you're ready for a senior role.

3.4 *Time Management*

Managing time effectively is one of the common traits among senior software engineers, and it's also one of the hardest skills. As you write more code and build up more domain knowledge, you'll start to become an integral part of your team. You'll become the expert on certain features and areas of the codebase, and people will come to you with questions about how something works or if it's possible to extend something you wrote with new functionality. You'll spend more time tracking down bugs, planning new projects, building out feature specifications, and possibly even helping to interview candidates to join your team.

Time management gets harder the more senior you get. Your team will become increasingly reliant on you to keep things moving, so it's important to get things done while not wasting your time.

As a junior engineer, it's good to be generous with your time, because anything you work on will help you learn. But as you prepare for a senior role, you'll need to learn to be careful with your time because it's your most limited resource. Senior engineers know how to work smarter, not harder, because they know that using their time wisely gives them leverage.

- They may spend a day or a week planning, without writing a single line of code, because they know that getting the solution right on paper will save them weeks of refactoring later in the project if the design needs to be changed.
- They know when to allocate their time to helping others, because sometimes helping unblock a coworker raises the entire team's throughput.
- They know when it's important to document a critical piece of the system, even if it takes them hours to type it all up, because it's worth spending the time to share the knowledge.
- They know how to find chunks of time in their day for deep focus and writing code, even if that means declining meetings or blocking off time on their calendar.

- They timebox their work. They know that it's better to ask for help or move to a different task than to spend too much time running in circles.

The best programmers understand how important their time is and will optimize their tasks and focus on those that will have the greatest impact based on the time constraints.

3.5 *Attention to Detail*

Programming is naturally a detail-oriented task. Take any programming language's syntax for example—you could have a codebase with hundreds of thousands of lines performing complex data processing, but if you misplace one semicolon or forget to close a parenthesis somewhere, your program will grind to a halt. Even worse, your logic may be flawed even though the syntax is correct, which will lead to frustrating nights trying to figure out why your program compiles but doesn't behave the way you expect it to.

Almost every aspect of delivering software requires focusing on details—things like defining requirements, reading other people's code, writing correct logic, implementing thorough error handling, and analyzing logs and other structured data.

As professional programmers, we're expected to prototype solutions and write applications to solve customer pain points. But it's not enough to focus only on the default use cases and ignore error handling or how our users interact with our programs. Good engineers try to anticipate every way in which a program can fail, and then work to put safeguards in place to prevent those scenarios from happening.

And they're not just detailed in the code they write. They bring attention to detail throughout the software development process. They pick apart the technical requirements and clarify any ambiguity, because they know that well-defined specifications aid them when it comes time to handle edge cases in their code.

How are senior software engineers detail-oriented?

- They have an ability to find gaps in the requirements and ask the right questions to fill those gaps.

- They tend to be diligent about keeping their codebase clean and organized.
- They read other people's code, sometimes more than once, to understand what it's doing. Then, they try to figure out ways in which it can fail.
- They are meticulous about covering all the bases, whether it's through test coverage, input validation, or handling edge cases. The more scenarios their code handles, the less chance it has to fail.
- They know how important good documentation can be and will find time to write it, even if it's not the most glamorous part of the job.

High-quality software requires a high attention to detail, not just in the code but in the entire software development life cycle.

3.6 *Engineering Excellence*

It takes persistence, determination, and many years in order to perfect a craft, but in the case of Jiro Ono, a sushi chef made famous by the documentary film *Jiro Dreams of Sushi*, he spent his entire life in pursuit of perfection.

Always looking to improve, Jiro worked hard every day to improve all aspects of his craft, from sourcing better ingredients, to preparing his dishes, to delighting his customers with the highest quality sushi. He was so determined to deliver the best experience possible that he fixated on every aspect of the meal, even changing the orientation of the sushi on the plate if the customer was right- or left-handed.

These small improvements compounded over the years, and Jiro's small 10-seat restaurant located in a Tokyo subway station, Sukiyabashi Jiro, became the first sushi restaurant in the world to receive three Michelin stars.

Although programming requires a very different set of skills than making sushi, writing software is also viewed as a craft, and as programmers, we are always looking to improve our coding skills.

Engineering excellence is a means to an end, not an end goal in itself. It is the relentless pursuit to raise the bar in terms of the quality of our software and the speed at which we can deliver it to users. Senior engineers continuously strive towards engineering excellence to improve the processes by which their team delivers software. They identify barriers and obstacles that prevent themselves and their team from doing their best work, and then work to remove those barriers in order to unlock higher quality and greater throughput for the team as a whole. In doing so, the goal is to delight users with the best experience possible.

3.7 *Team-First Approach*

There's a lot that has been written about teamwork, and for good reason. When you have a group of people all working towards a common goal, you can achieve great things. Each individual member on the team brings with them a unique set of skills, and when a team is able to leverage the skills from one of their team members, everyone benefits.

Senior software engineers recognize that shipping software is a team sport. While programming may feel like an individual activity when you're deep in the code, developers rely on one another to review their code, answer questions, share knowledge, and teach each other.

When software developers work together, they can achieve so much more than what each individual could accomplish on their own. When they build off each other's work, fix each other's bugs, and share their knowledge, a team becomes greater than the sum of its parts.

A senior engineer recognizes the importance of the team, and they identify and complete tasks that benefit the team, even if the work isn't planned or assigned to them. They understand that sometimes being senior means taking on the mundane tasks that no one else wants to do, because it will help unblock others to complete more work. Their focus is always a team-first approach, and they strive to lift up all members of their team because they know that when it comes to software, we all share the same responsibility to keep the systems running and deliver as much value to our users as possible.

Senior engineers recognize that the team consists of more people than just other software engineers. Test engineers, product and project managers, product owners, scrum masters, designers, and technical writers can all be a part of the same team or business unit. Just because someone doesn't write code doesn't mean they're not part of the team, and senior engineers work with everyone to achieve the team's goals.

Senior engineers look for opportunities to mentor junior engineers, help them develop their coding skills, and teach them problem-solving techniques. We all start out as junior engineers when we begin our careers, and senior engineers are able to empathize with the younger members of the team and help them work through problems collaboratively when they notice someone is stuck or unsure of which direction to go.

And finally, senior engineers understand that in some cases, sacrificing their own productivity in order to unblock or enable other developers' productivity is time well spent. They may spend a day without coding while they work through requirements for a project because they know it will enable the rest of their team to work quickly with clear instructions. They view productivity in terms of the whole team, rather than just themselves, and sometimes that means working through others or spending more time upfront in order to enable their team members to move faster in the future.

3.8 *Job Level Matrix*

As companies hire, grow, and promote from within, they often need to make decisions about the requirements and responsibilities for each role, including each individual's salary compensation. Job levels are a tool that many organizations use to standardize these decisions and explicitly define and document the responsibility level and expectations for each role at a company. Job levels are typically associated with pay bands or salary ranges for each level, and different companies structure their levels differently depending on their unique organizational needs.

So, why do companies utilize job levels?

Job levels allow organizations to be strategic with their hiring decisions and bring more consistency to the hiring process. Levels provide a helpful framework for how a company should hire, promote, and retain their talent, as well as providing a way to highlight specific areas for improvement, while developing their employees along their career growth trajectories. Additionally, job levels bring fairness and transparency to promotion decisions by providing a structured approach to identifying when an employee is ready to move to the next level with a promotion.

So, how can you use your company's job levels to leverage a promotion?

Whether you like it or not, climbing the corporate ladder is a game. And the job levels provide an even playing field for you and your peers.

Fortunately, these job levels also lay out the rules of the game for everyone to see, so the better you understand the rules of the game, the better you can use them to your advantage.

First off, if your manager or HR department hasn't already provided you with your company's job level matrix, you should ask for one. Most larger companies should have a job-leveling framework in place, but not all do. If you're working at a smaller company or a startup, they may not have one, but it's still good to ask. At the very least, it will show your manager that you're motivated to grow your career and work your way up the org chart, and you might even convince them to create one.

Once you have a job level matrix, you'll be able to see exactly what is expected of you in your current role. You'll be able to see your current responsibilities and identify areas that you know you need to improve. Even better, you can also see what skills and responsibilities are expected for more senior roles, and it provides a blueprint for exactly what you should be working on in order to jump to the next level.

So, let's look at what a common job level matrix looks like for programmers. It's valuable to view a junior role and a senior role side-by-side so you can compare and contrast the differences in impact and responsibilities. Although it may already be apparent, it's important to note that there is an assumption that these levels are cumulative, meaning that a senior programmer is expected to meet all the criteria in their current level, in addition to all the criteria and responsibilities in the lower levels as well. Keep this in mind as you're working towards your promotion, because you need to build a solid foundation in your current level before you can work on the responsibilities of the next level.

Here's what a common job level matrix for junior and senior programmers looks like.

	JUNIOR PROGRAMMER	**SENIOR PROGRAMMER**
Job Scope	Performs moderately complex problem solving with guidance and assistance. Is able to follow functional specifications for new features. Primarily involved in implementing systems that others have designed. Expected to spend majority of time learning the system and software development best practices.	Performs complex programming tasks, often without guidance or assistance. Has solid understanding of the impact of their code on the team's infrastructure. Has a strong understanding of how the different components of the system fit together. Is able to write thorough specifications for new features. Primarily involved in designing systems that they and others will implement. In-depth understanding of software development best practices.
Knowledge	Demonstrates solid understanding of programming fundamentals (data structures, object-oriented design, algorithm complexity). Has strong knowledge in their core area of the tech stack (programming language). Has general understanding of the components that make up the overall system.	Is highly competent in one or more areas of the tech stack (multiple programming languages). Can design a modular solution to medium and large problems. Possesses deep understanding of how the different components in the system interact with one another
Communication Skills	Communicates effectively in a number of scenarios: Code reviews, one-on-one conversations, small and large groups, management.	Is able to discuss technical trade-offs and alternatives in a rational and impartial manner. Has the ability to accept criticism of their designs and iterate based on feedback from others. Knows when to escalate issues.
Ownership	Contributes code that meets specifications. Troubleshoots issues and fixes defects in the codebase. Reviews code written by their peer. Contributes to documentation.	Identifies, proposes, and leads discussions that address problems impacting the team. Admits quickly to mistakes and works quickly to resolve them. Strives to learn from their own failures and failures of others.
Judgment	Offers their perspective during planning and design reviews. Proactively shares status updates with project stakeholders.	Is able to accurately estimate the amount of effort or complexity required by each task. Plans out work to ensure tasks are delivered on schedule. Proactively raises concerns with project stakeholders to ensure projects are delivered on schedule.

	JUNIOR PROGRAMMER	SENIOR PROGRAMMER
Character	Demonstrates initiative and an eagerness to learn. May provide mentorship to interns and peers.	Demonstrates initiative and a willingness to mentor others without being asked. Offers assistance outside of their own tasks when needed. Provides guidance to interns and junior level developers. Delivers constructive criticism when asked for feedback.

Table: Job level matrix for programmers.

This is meant to serve as an example job level matrix that you would find in any engineering organization. Your company's job levels may differ from the ones presented here, but the general idea will be the same. Your job as a developer is to identify which areas you feel you meet the requirements for, and which skills you know you are lacking. Once you've identified each category and where you stand, you have a blueprint for exactly what you need to work on in order to jump to the next level.

You may feel like you already possess certain skills or meet the requirements for the senior developer level, and that's great! But remember that to be eligible for a promotion from one level to the next, you must demonstrate a proven track record of meeting all the requirements in the job level for your current role, plus some or all of the requirements in the next level. Only once you've proven you can handle the responsibilities of the next role will you be considered for a promotion to a senior engineer.

And remember, it doesn't always have to do with how long you've been coding or how long you've been with your employer. The job levels are meant to provide an even playing field, and it's entirely possible that someone with fewer years of experience will get promoted before someone who has been working at the company and coding longer.

Additionally, getting promoted to a senior developer role is not just about a change in your title. As you can see in the job level matrix, moving up to a higher level involves a broader set of responsibilities and more trust that you are capable of solving complex problems and that you always put your team's interests before your own. In many cases, this also means there is more pressure for you to perform and deliver results on time. A senior role brings increased accountability along with more ambiguity in how to achieve your expected outcomes.

When it comes time for your annual review, it'll be significantly easier to ask for a promotion to a senior role if you're able to demonstrate that you consistently meet all the requirements in the job level matrix. One of the first things your manager will do is compare your past performance to the traits laid out in the senior job level matrix. And any requirements you haven't met will immediately jump out as you're being considered for the promotion.

Hopefully, you've learned enough to put yourself on the path to success. You should now have a better understanding of what you need in order to work towards that senior role, but it's important to realize that it won't happen overnight. The journey to becoming a senior developer takes years and requires a tremendous amount of discipline, hard work, and most importantly, constant learning. You are a student of the craft, and you will continue to learn throughout your career.

While becoming a senior programmer may be your goal, it should not be your destination. You should always be pushing yourself for continuous improvement. If anything, your learning will accelerate as you progress in your career, because you'll be building on the foundation you will lay in these next few years. You'll start to see how all the puzzle pieces fit together, and you'll begin to design, influence, and engineer complex systems. You're building a strong foundation today so that you can solve your customer's toughest problems in the years ahead.

🔗 RESOURCES

- The Differences Between a Junior, Mid-Level, and Senior Developer[10] (betterprogramming.pub)
- Programmer Competency Matrix[11] (sijinjoseph.com)
- Job Titles and Levels: What Every Software Engineer Needs to Know[12] (holloway.com)
- Software Engineering Levels[13] (dev.to/shavz)
- Software Engineer Qualification Levels: Junior, Middle, and Senior[14] (altexsoft.com)

---

10. https://betterprogramming.pub/
    the-differences-between-a-junior-mid-level-and-senior-developer-bb2cb2eb000d
11. https://www.sijinjoseph.com/programmer-competency-matrix/
12. https://www.holloway.com/s/trh-job-titles-levels-fundamentals-for-software-engineering
13. https://dev.to/shavz/software-engineering-levels-35p0
14. https://www.altexsoft.com/blog/business/
    software-engineer-qualification-levels-junior-middle-and-senior/

4 You're Not an Impostor

Have you ever sat in a room and felt like you weren't smart enough to be there? Maybe the conversation moved too quickly for you to follow, or perhaps your coworkers debated a topic you weren't familiar with. It's an uncomfortable feeling that every programmer experiences at some point in their career. When you start to doubt your own abilities and feel insufficient at your job, you're experiencing impostor feelings.

> ❯❯ EXAMPLE
>
> Feeling like an impostor can come on suddenly in a number of different situations:
>
> - You question your ability to complete a task you're stuck on.
> - You feel like you don't belong on a team or at a company because you're not capable enough to follow along during discussions.
> - You feel completely overwhelmed, especially when starting a new role, a new project, or learning a new codebase.
> - You have trouble articulating how something works on a technical level or when someone asks you why you made a certain decision.

If you've ever found yourself in one of these situations, it's hard *not* to feel like a phony. The good news is that it's probably not as bad as you think, and it's way more common than you realize. Nearly everyone has these feelings at some point. Unfortunately, part of the problem is due to the stigma surrounding impostor feelings, which often discourages people from discussing their feelings openly and seeking advice from others. It's especially hard during the first few years of your career because you're learning so many different things all at once:

- Your team's codebase and the technologies they use.
- Your team's development processes and methodologies.
- Your company's business logic and business model.
- Your manager's management style.
- How to work well with your team members.
- How to work with stakeholders, product managers, designers, and executives.

It's common for software developers to feel overwhelmed with the pace of things during the first few years of their career. Things move incredibly fast when you're still getting up to speed, and decisions happen quickly. Try not to get discouraged if you're having trouble keeping up with your coworkers as you're still learning. Some of your coworkers have been with your company for years, so they will naturally have built up more domain knowledge over time and gained a deeper understanding of the problems that need to be solved. They've probably written large portions of the codebase, so they will know the system better.

◇ IMPORTANT   The key thing to remember is that you shouldn't be intimidated by coworkers who know more than you. In fact, you should view it as an opportunity to learn from them and ask them questions.[§7] Leverage their knowledge so that you can learn faster from their experience.

4.1   *It's a Feeling, Not a Syndrome*

Chances are you've heard people talk about feeling like an impostor, although they probably referred to it as having "impostor syndrome." While this phrase is commonly used, it's misleading because it tends to oversimplify the issue, but it does help you easily communicate the feelings you're experiencing. It also does a good job of capturing a feeling of being overwhelmed or inadequate that you may be struggling with. While this may seem like a good thing, there are some negative consequences that come from using the term.

The solution is not to label yourself as having a medical syndrome that needs to be treated; it's to recognize your feelings so that you can work towards improving yourself and building confidence. Slapping a label on something you're feeling doesn't make it go away, but recognizing that feeling is the first step towards self-improvement.

Additionally, the term "impostor syndrome" tends to imply several things for some people:

- You either have it or you don't.
- It's abnormal.
- As with many medical syndromes, you likely won't get better without treatment.

None of these points are true, at least for most people. Think of it more as a natural *feeling* you have, like stage fright or feeling uncomfortable. It may come on suddenly and with varying intensity, but it's a temporary feeling that will eventually pass. It's good to remember this when you're having these feelings so you can manage them and try to reduce the accompanying negative thoughts.

On top of all that, what most people don't understand is that there's no officially recognized "impostor syndrome" among medical professionals. There is not a single mention of impostor syndrome in the *Diagnostic and Statistical Manual of Mental Disorders* (DSM), which provides a common language and standardized criteria for classifying mental disorders and is widely used by mental health professionals.

In fact, the term *impostor phenomenon* was first described by Dr. Pauline R. Clance and Dr. Suzanne A. Imes in 1978.[15] Clance and Imes first defined the impostor phenomenon as an "internal experience of intellectual phoniness," or, in other words, feeling like a fraud. While their study focused on high-achieving women who experienced these feelings, further research has revealed that both men and women are equally susceptible to the impostor phenomenon. Clance and Imes's study also showed that these feelings are prevalent in highly successful people and those that have worked hard to build their careers.

In essence, impostor feelings can happen to anyone and can come on at any point in one's career. Rather than trying to fight it, it's more effective to embrace it and use it to motivate yourself to improve. Let's dig deeper into what you can do when you suddenly start to feel like an impostor.

4.2 *Knowledge Gaps*

As programmers, we're expected to know a wide array of things in order to do our jobs. On top of fundamental skills like writing code that compiles without errors, we also need to know how to design our programs to be

---

15. Clance, Pauline R. and Suzanne A. Imes. "The Impostor Phenomenon in High Achieving Women: Dynamics and Therapeutic Intervention." *Psychotherapy: Theory, Research & Practice* 15, no. 3 (Fall 1978): 241–247. [16]

16. https://mpowir.org/wp-content/uploads/2010/02/Download-IP-in-High-Achieving-Women.pdf

future proof, how to write automated tests so we can be sure the code works as expected, and how to deliver results consistently and efficiently.

It's easy to feel overwhelmed with the amount of knowledge required in order to complete our day-to-day tasks, in addition to answering questions from project stakeholders, defending our technical decisions during code reviews, and planning for the future.

There will be plenty of times throughout your career where you won't know the answer to a question or won't know what someone else on your team is talking about. It happens to software engineers of all levels, not just junior engineers. The thing that makes good engineers stand out from the rest is that they are humble and aware that they don't know everything, and they use that knowledge to learn and grow.

> ❱ EXAMPLE

Situations that might bring on impostor feelings include:

- Your boss might ask you a question in front of your team that you don't know the answer to.
- A senior engineer might ask if you considered an obvious edge case when fixing a bug.
- A senior product manager might ask you to estimate how long it would take to build a new feature in a part of the codebase you're not familiar with.
- Your boss might ask you to clarify how something you built six months ago works, and you might not remember all the details about it.

It's okay to admit that you don't know something, or that you hadn't considered some obvious edge case. In fact, admitting that you don't know something is a sign of maturity. If you want to be successful in your career, you need to put your ego aside and accept the fact that you won't know everything. It's nearly impossible to know everything there is about programming. It's such a broad industry that is still evolving each and every day, and it's simply not possible to be an expert on every topic.

Additionally, there's even more to learn about the *process* of delivering software, because writing code is only one part of the equation when it comes to doing our jobs. There are planning, analysis, design, implementation, testing and integration, and maintenance phases in a modern software development life cycle, and each phase comes with its own nuances and best practices. While you may not be great at all of those phases, a senior engineer must understand the entire process. It's normal to feel like an impostor if you don't have a lot of knowledge in one or more of these areas, but that doesn't mean you're not a good engineer.

Let's look at what you can do to minimize those impostor feelings and leverage them to become a better software engineer.

4.2.1 IDENTIFY THE GAPS

You have gaps in your knowledge. We all do. It's natural, so don't worry about it too much. What's more important is that you seek out and acknowledge these shortcomings so that you can take steps to close the gaps. Accepting your limitations allows you to put your ego aside and open your mind to new ideas and opportunities to grow.

Too often we fall into the misconception that we know more than we really do, which leads us to be stubborn and closed-minded when we're presented with new information. Believing you know more than you do can have a negative effect on your ability to learn, and in some cases, it can stagnate your career growth. It can also make it easy to get defensive when your ideas are challenged. How you react in those situations can lead to conflicts with your coworkers or hinder your ability to make impartial decisions, which is why it's important to be honest with yourself about the limitations of your knowledge and your skills.

When you accept that you don't know what you don't know, it opens the door to opportunities to improve the situation:

- When you're honest with yourself about what you don't know, you naturally identify areas where you need to seek more information and learn more.
- When you accept you don't know something about a topic, it takes the pressure off when you need to ask for help. It gives you confidence to reach out to someone else who may be an expert in that topic and can help you, or they may be able to point you to someone who can.
- Knowing what you don't know helps prevent you from making an uninformed or biased decision.

So, how do you identify what you don't know? A good first step is to write down in your notes things you already *know* that you don't know. This could be things like topics, phrases, or acronyms that you've heard before but don't know what they are. Write down as many things as you can think of.

◇ IMPORTANT Use your favorite note-taking app or even a physical note-book: something that you'll be able to access at any time and be able to reference later on.

It's important to be honest with yourself during this step. Don't be embarrassed if the list is long or you think it's full of basic topics—you don't need to share it with anyone. And these don't necessarily have to be technical topics either. They can be related to business in general, some unique aspect of your company, or something industry-specific. And finally, while it's important to write down things you don't know or understand, it's also important to write down *what* you don't understand about it too. This will give you some indication on what you should focus on when it's time to learn.

❯ EXAMPLE

- I don't understand how containers are different from VMs. It sounds like they do the same thing.
- I don't understand dependency injection or why I would need it.
- I don't understand our company's sales process. Where do our customers come from and why do they want our product?
- I don't understand React hooks. When would I even use them?
- I don't fully understand the difference between an abstract class and an interface, and when to use one over the other.
- I don't understand the difference between stack vs. heap memory.
- I don't understand the economics of our industry. How do we determine the price point to sell our product competitively?

The next step is to observe. Take note of any new terms, phrases, acronyms, or concepts that pop up during conversations, video calls, chats, or comments from your coworkers or things you come across online. Anytime you come across an idea that you're not familiar with, add it to your list. And every time you hear someone mention a topic that's already on your list, add a checkmark or a +1 next to it.

- What is ARR? And why is the leadership team interested in tracking that metric?
- My boss keeps mentioning the need for static code analysis. What is it and why is it so important?
- I keep seeing people mention Kubernetes. I know it has to do with containers but why is it so popular?

Before you know it, you'll have a long list of topics of things to learn about. Whenever you have some time to spare, you can use this list as a starting point for things to research.

4.2.2 CLOSE THE GAPS

Once you've compiled a list of topics that you know you don't know, you can begin to work towards learning and closing those knowledge gaps. At this point, you've essentially compiled a roadmap for expanding your knowledge. Now, you can begin to chip away at this list and start to close those gaps in your knowledge.

You can start with any topic, whether it's something you're really interested in or something you think will help you during your career. Another option is to rank the topics by what you think is the most important or what will have the highest impact in your day-to-day work.

An important thing to remember during this step is that you shouldn't try to learn everything at once. Your list may be long, but that's okay! Focus on learning one topic at a time so you can give it your full attention. Try to learn as much about it as possible before moving on to the next topic. If you try to learn too many things at once, you won't be able to give each topic the attention it deserves, and you may not remember each thing you're trying to learn.

You may be able to learn about some topics in a day, while others may take days or weeks to fully comprehend. Feel free to dive as deep into a topic as you see fit, but always try to learn more than just surface-level details about something.

Let's look at the ARR acronym from the previous example.

Through your research, you learn that "ARR" stands for annual recurring revenue. Great! Now you'll know what the executive team means whenever they say "we need to focus on growing ARR." But don't stop there—try to dig further and understand why focusing on growing ARR is so important.

- Why is ARR an important metric for a SaaS company?
- How is ARR calculated? (knowing this will help you understand how to increase it)
- What does it mean if ARR is growing? What about declining?
- How does my work fit into the company's goal of growing ARR?

Spending time to do your own research and teaching yourself a topic is invaluable. Try to learn the foundational concepts behind each topic, and what makes them important. Once you start to learn about a topic beyond just the basics, you may come up with more questions and want to dig deeper. Use this as a good opportunity to reach out to others and ask some questions. Try to identify coworkers or people in your network that are familiar with the topic and let them know you'd like to learn more.

🔥 CONFUSION   Sometimes, the hardest part is in determining exactly what you should ask, who you should ask, or how you should frame your questions. Asking questions is harder than it sounds, so we'll cover that topic in greater detail in How to Ask Better Questions.[§7]

Identifying and closing your knowledge gaps is just one thing you can do in order to counteract negative thoughts when you're feeling like an impostor. It's a great way to build confidence, expand your knowledge, and learn new skills and ideas that could help you throughout your career. Next time you're feeling like an impostor, try writing down things you don't know or want to learn, and start chipping away at them to learn something new.

🔗 RESOURCES

- Learning at Work[17] (jvns.ca)

---

17. https://jvns.ca/blog/2017/08/06/learning-at-work/

4.3 *Dealing with Criticism*

We've all been criticized at some point in our careers, and it's safe to say that it's never a great feeling. It can be difficult to receive criticism, even if it's intended to be constructive. Criticism can be a hit to your ego and sometimes leave you feeling lost and confused, especially when you feel so confident in your work.

Whether you like it or not, criticism is sometimes necessary in order to foster a healthy engineering organization and to maintain an organized codebase that can grow and evolve over time. As a developer, you'll receive feedback throughout your career in many different areas, such as:

- **Code reviews.** Other engineers may find issues with the syntax, logic, or readability of the code you submit for review.
- **Design reviews or brainstorming sessions.** There is always subjectivity when it comes to architecting solutions to solve problems. Sometimes your proposed design will differ from your coworkers' ideas, or they may pick apart your design.
- **Your manager.** You may receive direct feedback from your manager about specific things that you need to improve, such as your organizational skills, attention to detail, work ethic, ownership, or attitude.
- **Product managers.** Although you may have spent weeks on a feature, the solution you deliver may not be in line with what the product owner or customer was expecting.

While no one ever wants to be criticized, how you process the feedback can make a big difference. Accepting constructive criticism at work is an important stepping stone towards developing maturity and becoming a better developer. The hard part though is to learn to separate the content of the suggestion from the way that it was delivered. Criticism is not always delivered well, so it's easy to get upset if it's done in an insensitive way. And if the criticism is delivered over chat, email, or some other form of written text, the tone is often lost and can be interpreted differently than the author intended it to be.

The thing to keep in mind is that you can often still learn a great deal from criticism, even if it's delivered in a way that's insensitive or not as constructive as it could have been. If you are able to mentally separate the essential content of the criticism from the style or tone of the feedback, you'll still be able to learn from it, even though it may not be what you wanted to hear.

Additionally, it's important to separate the criticism from the person giving it. There may be coworkers who you don't always agree with, that get on your nerves, or that have rubbed you the wrong way in the past, but that doesn't mean you should automatically discount their feedback. You could learn some valuable things from these people, if you're able to let go of any biases you may have towards them.

So, what can you do in these scenarios? Let's look at a few things you can do to turn a negative situation into a positive one.

4.3.1 CONTROL YOUR RESPONSE

This is the most important and often the most difficult aspect of dealing with criticism. The way you respond to criticism is an indicator of your maturity and professionalism as a software engineer. Your reaction may be visible to your manager and your teammates, and if you're not able to control your emotions when responding to criticism, you may risk hurting your reputation or your teammates' trust in you.

The first step is to be aware of your emotional reaction. If you notice you're getting upset about someone or something, ask yourself why it's making you so upset. Having self-awareness about your emotional state helps you process your feelings and be more mindful about how you should respond.

If possible, wait to respond until you've been able to reflect on the feedback and gather your thoughts. It's often best to refrain from responding immediately to feedback because your emotions may get the best of you. You may say something you will regret later, or you may not be able to think clearly in the moment. Instead, it's better to accept the feedback and allow yourself to process and internalize the suggestions first. Giving yourself some time allows you to reflect and think about how you should handle the suggestions instead of responding immediately with excuses or defensiveness, and you will avoid saying something you may later regret.

4.3.2 PROCESS THE FEEDBACK

No matter how good you are at your job, there is always room for improvement. When faced with constructive criticism from your boss or coworkers, try to reflect on the reason behind their feedback. Why are they giving you this feedback? There may be a good reason behind their actions, and if you're able to understand fundamentally why you're receiving criticism, you're more likely to learn from their suggestions and make better decisions in the future. There may be important details you've overlooked that they're bringing to your attention.

Try asking yourself these questions when you receive constructive criticism:

- Are there reasons why I might have approached the problem in the wrong way?
- Where does my design fall short?
- Did I misinterpret the problem or task?
- If there is a better solution proposed, what makes it better than mine?
- Are there solutions that combine benefits of my approach with a suggested alternative?

◇ CAUTION Note that it's never 100% certain that the feedback you're getting is correct and that your approach was wrong. Sometimes, developers who review code don't have the full context or may misinterpret something.

Asking yourself these questions, along with others designed to get to the root of the issue, will help you better understand the downsides of your approach. Perhaps there was additional context that would have been helpful to know when making a decision, or maybe there was a design pattern you weren't aware of that was a better fit for the problem. Whatever the case may be, it's important you understand why you're receiving the feedback in the first place. This self-awareness alone will show that you are mature enough to handle criticism in the first place.

4.3.3 LEARN FROM OTHERS

Receiving unexpected feedback may hurt your confidence, but don't lose sight of the bigger picture. This is your opportunity to grow as an engineer. Once you've accepted that you can be better, the hard part is over. Then, it's time to figure out what you can do to improve. These small changes compound over time and can help you become a better engineer quicker than you think.

There's a lesson to learn every time you receive constructive feedback, so don't squander an opportunity to improve yourself. If you choose to ignore the feedback, you could be missing opportunities to grow as an engineer, especially if it comes from people with more seniority than you. They may have years' worth of experience that they're trying to pass on to you. Perhaps they made a mistake when they were younger, and they don't want to see you repeat those same mistakes. Or maybe they've noticed something that you could be doing better that helped them earlier in their career. Whatever the reason, don't take advice from your senior coworkers for granted. They've been in your shoes. They're trying to help you avoid the same mistakes they once made themselves.

> ⌗ **RESOURCES**
>
> - How to Make Your Code Reviewer Fall in Love with You[18] (mtlynch.io)

4.4 *Track Your Accomplishments*

As developers, we have to maintain incredibly complex mental models about how codebases work, and those models change continuously as new code is written, merged, and deployed to production. Over time it becomes harder and harder to recall what you worked on. When you think back on what you've built over the last few months, it may be difficult to remember everything you worked on. Days, weeks, and months go by, and you may not feel like you've accomplished a whole lot, which sometimes leads to impostor feelings. In reality, you complete a lot more work than you probably realize. The hard part is remembering everything you've done.

A simple but effective way to curb those impostor feelings is to keep a log of what you've worked on each week. Keeping track of what you've accomplished has a number of benefits:

- If you're ever feeling like an impostor, it helps to look back at all the features you've built, bugs you fixed, and your major accomplishments throughout your career.
- When it comes time for your quarterly or annual review, you'll already have a list of accomplishments you can refer to when asking for a promotion or a raise.

---

18. https://mtlynch.io/code-review-love/

- If you find yourself looking for new employment opportunities, you can refer to this list when updating your resume.

All it takes is a few minutes every week to write down what you worked on, and you'll benefit down the road when it's needed.

It's never too late to start logging your accomplishments, and if you can't remember what you worked on, you can always go back to your project management system and filter for the closed tickets that were assigned to you. There may be other things you've done that weren't logged in your ticketing system, but it's at least a good place to look if you're just getting started.

>> EXAMPLE

So what things should you track?

- Any goals or OKRs (objectives and key results) you reached and how you reached them, along with facts, quantifiable analytics, or financial data points to back it up. Basically, anything you improved that has a number, percentage, or dollar amount attached to it.

  - "I helped increase conversion rate by 1.2% last quarter by fixing multiple UX issues on our checkout flow."
  - "I was able to reduce customer service calls by 19% by building out a self-service knowledge site for our customers."

- Difficult situations or challenges with coworkers, customers, or third parties that you navigated successfully. Take note of the path you took towards a resolution.

  - "I identified and fixed a critical bug that caused downtime for one of our largest customers."
  - "I helped our team reach an agreement on a new database schema by identifying areas where one design scaled better than the other."

- Tasks or projects that you completed on time or ahead of schedule.

  - "I completed the rollout of our new automated invoicing system two weeks ahead of schedule."

- Take note of times when you exceeded expectations.

  - "During my free time, I cleaned up our internal documentation on our team wiki, removing outdated pages, adding missing sections, and including diagrams to visualize how our systems fit together."

When you take the time to write down your accomplishments, you'll be able to reference them any time. It's a good habit at the end of every week, month, or team sprint. It only takes a few minutes each time, and it helps you focus on your wins rather than your gaps in skills or knowledge. And remember, anytime you're feeling like an impostor, you can always come back to read through everything you've accomplished.

Hopefully, you now understand that feeling like a phony or a fraud is common among all software developers, and that you shouldn't get too hard on yourself when you're going through a period of feeling like an impostor. The most important part is to try to identify *why* you're experiencing those feelings and accept that nobody is perfect. Once you understand the root cause of why you're feeling a certain way, you can begin to take action to improve yourself and your skill.

With a little persistence and determination to get better, you'll be able to build the confidence to take on any situation, no matter how daunting.

🔗 RESOURCES

- Actual impostors don't get impostor syndrome[19] (zapier.com)
- I'm An Impostor[20] (dev.to/bytebodger)
- How to Live with Chronic Imposter Syndrome[21] (eugeneyan.com)
- Dealing with Imposter Syndrome[22] (dev.to/jingjing142)

---

19. https://zapier.com/blog/actual-impostors-dont-get-impostor-syndrome/
20. https://dev.to/bytebodger/i-m-an-impostor-5f7f
21. https://eugeneyan.com/writing/imposter-syndrome/
22. https://dev.to/jingjing142/dealing-with-imposter-syndrome-449i

5 Working with Your Manager

Your professional relationship with your manager is one of the most significant factors affecting your career growth. You are ultimately responsible for your own career, but your manager has more influence over your career trajectory than anyone else in your company, so it's critical that the two of you work well together. They sit above you in the organizational hierarchy, which gives them power to make decisions that can affect your career, both positively and negatively. This relationship has far-reaching consequences, such as which projects you work on and the opportunities you're given. Many say that your relationship with your manager is the single biggest factor that affects your job satisfaction.

Some managers are great to work with, but others may prove more difficult. No matter how good or bad your manager is, it's your job to make the relationship work. Every professional relationship between a manager and a programmer is unique, and you'll have to work hard to figure out what works best between the two of you. Your relationship with your manager will evolve over time, and you'll have to continue to adjust and make changes to how the two of you work together.

5.1 *Build Trust*

Trust is an important foundation for working well with your manager. If you can't trust each other, your relationship with your manager will break down and your job will be more difficult than it needs to be.

If your manager can't trust you can do the job they ask you to do, you'll miss out on projects and opportunities to grow and be promoted. If they can't trust you'll communicate openly and honestly about the status of a project, they'll have no choice but to micromanage you to get the information they need.

So, how do you build trust with your boss?

The first step may be obvious, but it needs to be said—do your job. There's no way around this. It's table stakes for everything else discussed in this section. If you don't do your job, you're going to have a very difficult time building trust with your manager, so that should be your number one priority. It's literally just doing what is asked of you according to your job description.

The next step is to do your job *reliably*. Your manager needs to know you are reliable and that they can count on you to complete the tasks you are assigned. Doing your job reliably does not mean you won't make mistakes, your estimates will always spot on, or your projects will always be completed ahead of time. Nobody is perfect and timelines will slip, bugs will happen, and estimates will be wrong. That's just a part of building software that you need to learn to live with. Being reliable means that your boss can count on you to take on a task and see it through to the finish line. The more often you do this, the more your boss will know that you can take on whatever they throw your way.

There's more to being reliable than just completing your tasks. Here are some other examples of what it means to be someone your manager can rely on:

A reliable software engineer is someone who is predictable (that is, does what they say they will do) and who is able to complete their tasks from start to finish. Doing just these two things will help develop a good foundation of trust between you and your manager.

5.1.1 UNDERSTAND YOUR MANAGER'S GOALS

While your manager expects you to deliver results each quarter, they have their own goals, milestones, objectives, and key results that they need to deliver as well. They are responsible for all the work your team delivers. Understanding this point will help you better understand some of the decisions your manager makes.

As a software engineer, your job is to do everything you can to support your manager so they achieve their desired outcomes. Therefore, your manager's goals are your own goals.

◇ IMPORTANT This is an important concept that not many engineers understand early in their careers, because most of them are primarily concerned with proving their technical abilities.

Your work is just one piece of the puzzle that your manager needs to solve. When you understand how your work fits into the bigger picture, you'll be able to identify which tasks will help your manager reach their goals and prioritize those first. If you can manage to do this, you'll almost certainly gain your manager's trust. Conversely, if you hinder your manager's ability to meet their objectives, you may lose the trust of your boss.

Your boss has a goal this quarter to upgrade the programming language for your legacy codebase to the next version. They need to do this because the version you're on is no longer receiving security updates and there are new features in the next version that your team can take advantage of during development.

In the bad example above, you may think that you're demonstrating your growing technical skills and showing your manager that you're a good engineer who can solve difficult problems. The issue is that the tasks you worked on were not aligned with your manager's priorities for the quarter, whereas in the good example, you helped your boss get closer to their goal of upgrading the programming language version. If you want to build trust with your boss, you need to figure out how to align your own goals with your manager's goals.

⌀ RESOURCES

- How to Earn Your Manager's Respect[23] (hbr.org)
- How to Effectively Talk to Your Boss: 25 Dos and Don'ts[24] (careeraddict.com)
- Things your manager might not know[25] (jvns.ca)

5.2 *Adapt to Their Management Style*

Everyone works differently. There's no single way to maximize your productivity that works for everyone, and everyone has their own way that works for them when they need to get things done. This is *especially* true when it comes to managing people and projects.

---

23. https://hbr.org/2016/12/how-to-earn-your-managers-respect
24. https://www.careeraddict.com/talk-to-boss
25. https://jvns.ca/blog/things-your-manager-might-not-know/

Different managers have different management styles, so when it comes to working well with your manager, you'll need to figure out what style they prefer. You'll need to consider the following questions when determining how your boss prefers to work:

- How do they prefer to communicate?

  - Email
  - Asynchronous chat
  - Voice or video chat
  - Face-to-face conversations

- How often do they expect updates from you?

  - Once a day
  - Once a week
  - As needed

- What time of day are they usually available?

  - Mornings
  - Midday
  - Evening

- Do they prefer if you provide them with objective analysis for each option, or do they prefer to hear your personal judgment when making difficult decisions?

Although these aren't things you need to think about often, knowing the answers to these questions can make your life easier during high-stress situations like production incidents or tight deadlines. When you understand your manager's preferred way to work, you're less likely to make costly errors due to miscommunication.

Similar to how trust is a two-way street, a healthy relationship with your manager takes two people to make work. Treat your relationship with your boss as a partnership—you both share a responsibility to make it work. Unfortunately, part of that working relationship is out of your control; you can't control your boss, after all, but at least you can do your part to uphold your end of the deal. As long as you're making an honest effort to adapt to your manager's work style, they can't hold that against you when it comes time to conduct your performance review. However, if you don't make any effort to work well with your boss, they may use that against you.

So, what do you do if you're not sure what your manager's management style is?

Ask them.

It may be awkward, especially if you've been working with them for a while now, but it's an important conversation to have, and being both aligned on what you should be doing will help you in the long run.

During your next one-on-one, ask them how you two can work better together. It's important to be open and honest here, even though it may be a bit embarrassing because it's such a trivial question to ask. If possible, try to communicate your preferred style so they get an idea of how they can work better with you.

Honesty will always contribute to a better working relationship with your boss. The conversations may be uncomfortable or difficult, but they are worth it to build trust and respect. Your ultimate goal is to find the right balance when it comes to communicating with your manager.

Think of communication as a spectrum—one where you want to avoid the two extremes:

- **Under-communication:** Solving problems completely by yourself and not even telling your boss.
- **Over-communication:** Asking for help, guidance, or approval on every detail.

Neither extreme is right, and they will hinder your ability to build trust with your manager. The goal is to balance how much time and attention you demand from your manager when giving visibility or getting help, direction, or input.

It's a balancing act, and as a junior or mid-level software engineer, it's your job to figure out how to make it work. At the very least, try to build a habit of asking your manager for help only *after* you have at least one potential solution. This shows that you're not just dumping the problem in their lap, but instead asking for their help in refining the potential solution you already have. If you ask for their assistance empty-handed, it signals to your manager that you're not putting in the work before coming to them for help.

5.3 *Learn to Manage Up*

Having a healthy relationship with your boss makes your job easier, but there will be times when the two of you aren't on the same page. If your boss is overcommitted, overwhelmed, or even if they're not the best in a certain area of expertise, you need to learn how to manage up in order to make the relationship work.

To start, you need to recognize the situation you're dealing with. Perhaps you're dealing with:

- A boss that has been at the company for years while you're just starting.
- A boss that has just started while you've been with the company for a while.
- A manager who is a know-it-all.
- A manager that is new to the industry and may not know a lot.
- An indecisive manager.
- A manager that goes with their gut feeling instead of relying on data or the opinions of their team.
- A first-time manager that is still learning.

◇ CAUTION Before we dig deeper into how you can manage up, you first need to understand what managing up does *not* mean:

- Manipulating or deceiving your boss.
- Covering up a mistake you made.
- Hiding information from your manager that makes you look bad.
- Inserting yourself into office politics.

When applied correctly in the right situation, managing up can help you achieve the outcomes you're looking for, but if used incorrectly, aggressively, or in the wrong situation, it can backfire and hurt your image. It may take some time to learn how to effectively manage up, but when done correctly, you can get the results you're looking for.

As you pick up tickets, you may be able to complete some on time without any outside help, but other times you may run into issues that require you to seek help from your team members. This may be as simple as mentioning what you're stuck on during your team's stand-up and asking for help from someone familiar with the part of the codebase you're working on. Occasionally though, you may find yourself stuck due to external factors such as a dependency on another team or a technical reason why you can't continue.

5.3.1 USE LEVERAGE

As an engineer working towards a senior role, you should first try to figure out the issue on your own. Part of being a senior engineer is that you are able to complete small- to medium-sized tasks without any supervision from your manager. But sometimes you'll get stuck and need to bring in some help. Your boss may be able to help unblock you, and if not, they should be able to point you to someone who can.

In cases where you're stuck because of a dependency on another team within your company, your boss may have more leverage to ask the other team to do whatever you need them to do. Let's look at an example where you could leverage your manager's position in order to get what you need.

> ❯ EXAMPLE
>
> Suppose you're implementing a new feature for the sales team in order to simplify their workflow. If you don't have enough information to move forward on a feature request, reach out to the appropriate person on the other team to get your questions answered. If you've been blocked because you're waiting for a response, ask your boss if they can help get the answers you need. Sometimes, all it takes is getting your manager to contact the manager on another team to get your questions answered. Just be careful not to use this lever too much. You can earn a bad reputation with your manager if you escalate too often. It shows that you're not able to handle roadblocks on your own. Only use this as a last resort after you've done all that you can.

5.3.2 TELL THE TRUTH

Sometimes, you may need to inform your manager about something they may not be aware of. Your boss has to make many decisions each day, and sometimes, they may not have all the information they need. Speaking up in these cases is part of managing up.

Your boss needs to know that you have their back, and sometimes that means telling them things that they need to hear, even though they may not want to hear it.

> ❯ EXAMPLE

- Perhaps your company is planning a new feature to bring in some new business. Your manager may agree to take on a new project with a tight deadline without realizing there are technical limitations that will make that deadline impossible without taking on a lot of technical debt. You should let them know as soon as possible so the team can adjust the timeline as needed.
- You may have a boss that is new and isn't aware of a risk factor that could cause you to miss one of your quarterly goals, such as integrating with a third-party system. If the other party is dragging their feet and there's a risk of not hitting your deadline, let your boss know as soon as possible so they can manage expectations and modify the plan for the quarter.

It's better to have difficult conversations with your boss about something than to let it simmer and boil over. By then, it's already too late, and you'll have a high-stress situation on your hands. If possible, it's better to be open and honest with your manager so they can pivot or change directions if needed. In the end, they will appreciate the fact you gave them honest feedback.

5.3.3 MAKE YOUR WINS KNOWN

As individual contributors, we're deep in the codebase each week. We're naturally familiar not only with the inner workings and how different parts of the system fit together, but also with which parts of the system need work. It's easy for us to understand why a seemingly small bug is especially hard to fix, but it may not be apparent to someone who isn't writing code every day. You may deliver a feature that seems trivial but was actually a really challenging technical problem that needed to be solved.

Your manager's day will be filled with meetings, so they'll always be further away from the day-to-day technical challenges than they'd like to be. Your boss may not know all the details about the problems you're solving, so don't just assume your boss is aware of the exciting accomplishments you've made recently, or the technical challenges you've overcome.

Part of managing up is learning how to inform your boss about your accomplishments. This is especially important if you're close to or working towards a promotion. Ideally, your company will have a self-review process through which you can describe what major objectives you accomplished during each review period, but you don't have to wait until the performance review process to let your manager know how you're doing. Keep in mind your manager regularly gives progress updates to his boss, and he'll want to communicate about his team's wins and the progress they've made. To do that, your boss first has to know about what you've done. Try to find moments to let your manager know about your accomplishments, whether it's in private or public.

> ≫ EXAMPLE

Work with your manager to establish expectations on the types of outcomes and behaviors an engineer at the next level demonstrates, then find ways to let your boss know when you think you've demonstrated them.

> ⧉ RESOURCES

- What Everyone Should Know About Managing Up[26] (hbr.org)
- The Dos And Don'ts Of Managing Up[27] (idealist.org)
- How to Manage Up at Work[28] (wsj.com)
- 5 Tips To Manage Up At Work[29] (forbes.com)

5.4 *Dealing with Conflict*

You and your boss are both adults. You'll each have your own way of doing things, and you'll have your own opinions on how something should be done. Hopefully, you'll be able to figure out a way to work well together,

---

26. https://hbr.org/2015/01/what-everyone-should-know-about-managing-up

26. https://hbr.org/2015/01/what-everyone-should-know-about-managing-up

27. https://www.idealist.org/en/careers/managing-up

28. https://www.wsj.com/articles/what-does-it-mean-to-manage-up-11608242276

29. https://www.forbes.com/sites/carolinecastrillon/2020/02/23/5-tips-to-manage-up-at-work

but sometimes the two of you will have different opinions on how to accomplish a task.

If you have to disagree with your boss, do so politely and in private.

◇ CAUTION Do not surprise your manager with news in public. Doing so may catch them off guard and make them look unprepared in front of their colleagues, or even worse, their manager. It's possible your manager may interpret your actions as being disloyal to them.

5.4.1 CONTEXT BEFORE CONTENT

When you and your boss have a disagreement, they may get defensive because they may think you're challenging their goals. Oftentimes, they may focus on the intent of your actions, rather than the content of what you're discussing, which is why it's always good to clarify the context for why you're disagreeing with them. If you can manage to help your boss understand your perspective, they may be less defensive and more willing to see the argument from a new point of view.

If possible, try to frame your opinion in the context of a bigger goal or objective. Doing so will allow you to be more candid and honest when discussing your opinion, in addition to helping focus your manager's thoughts on the shared goal. It will also demonstrate that your difference of opinions is due to an external factor, rather than a personal attack on your boss's views. If you fail to provide context for your point of view, your manager may perceive your disagreement as a lack of commitment to their own interests.

5.4.2 RESPECT THE DECISION

You won't always be able to achieve the outcome you want, and in the end, your boss has the final say when it comes to important decisions that affect you and your team. If your manager considers your point of view only to decide against it, don't take it personally and don't hold it against them. It's better to respect their decision, be professional, and move on than to continue disagreeing with them. If it's not clear to you why they made the decision, consider bringing it up during a one-on-one. Perhaps there's another aspect to the problem that you're not seeing, and it will help to talk to your boss about the decision they had to make.

Part of becoming a senior software engineer means accepting that not every decision will go your way. Sometimes you need to let go of an opinion you are passionate about and move on. It's better to work together with your boss as a team and trust that they're making the right calls, rather than pushing back on every decision they make.

5.5 *One-on-Ones*

Recurring one-on-one meetings are your opportunity to receive direct feedback from your manager about how you can be better as a software engineer. A common misconception among junior software engineers is that one-on-ones are meant to give status updates on their current workload. The conversations with your manager during your one-on-ones should be about career growth, not your day-to-day work. Don't waste your opportunity by giving a status update about what you're currently working on. You should be talking about higher level things than individual tasks.

These meetings are just between you and your boss, no one else. It's precious time for you to be honest and talk about personal things. Try to avoid talking about things that can be discussed in the open with the rest of your team, because that's not a good use of your time during these meetings. Your one-on-one is a chance to talk about the difficult things that you wouldn't want to discuss in front of your teammates.

◇ IMPORTANT This can be awkward and uncomfortable at first, but the more open and honest you are about your feelings, the easier it gets.

Just be honest. This is your opportunity to get things off your chest. You have a direct and uninterrupted line of communication with your boss for a short period of time, so make the most of it.

While it's your manager's job to complete their team's long-term goals, they also need to fix processes and protocols that are broken or are not working for their team and their direct reports. They can't fix what they don't know is broken, however, so it's your job to be honest with them when something isn't working.

❯ EXAMPLE

- Let them know what challenges or frustrations you've had recently.
- Let them know if you're having trouble working with a difficult teammate.

- Let them know if a process isn't working and why.
- Let them know if you're feeling overwhelmed or burned out.

So, how do you make the most of your time during your one-on-one?

5.5.1 COME PREPARED

Set a meeting agenda ahead of time and make a point to discuss everything on the agenda. Add any topics you'd like to discuss or questions you may have for your manager. Setting an agenda ahead of time gives your manager time to prepare and get you the answers you're looking for. They may not always have an answer themself and may need to reach out to someone else for it.

Additionally, if you know ahead of time that they are going to ask you about a specific topic or task, make sure you have all the information you need in order to give them a sufficient answer. Your boss may set their own items on the agenda, so be sure to check it to see if there's anything that you need to prepare for.

5.5.2 TAKE NOTES

You should be looking for both positive and constructive feedback during these meetings, which means you should leave the meeting with concrete things you should be working on. Be sure to write these down during the meeting so you're able to reference them in the future. You have hundreds of decisions to make daily and multiple projects you're responsible for, so it's easy to forget specific things your manager asked you to do during your one-on-one conversations.

Plus, when you can look back on your notes, you can remind yourself of the things you need to work on. When your boss provides feedback, they expect you to listen and apply the feedback to your day-to-day work. Remember what you talked about, since they may bring up these areas of improvement during the next one-on-one. You want to be able to demonstrate you heard and reflected on the feedback.

5.5.3 ASK SPECIFIC QUESTIONS

This is your opportunity to ask for answers to specific questions you may have. "Specific" is the key word here. The goal here is to look for ways in which you can receive constructive feedback from your manager. This will help identify key areas you should focus on that will help you become a

better software engineer, or things your boss is looking for in order to help
you grow as an engineer.

> ❯ EXAMPLE

So what are specific questions you can ask?

- What are your top priorities right now and how can I help?

  - Remember that your job is to support your boss and help
    them achieve their goals. Asking them directly how you can
    help them reach their goals will help build trust, and they
    may give you specific tasks or projects that relate to their
    goals.

- What am I doing well that I should continue to do?

  - This question focuses on positive reinforcement for good
    habits and things that you're doing well. If you're doing a
    great job at something, you want to make sure you con-
    tinue doing so.

- What are some things I can improve?

  - You're looking for constructive feedback here. Your man-
    ager may give you specific things you can do to become a
    better engineer.
  - It's important that you make an honest effort to improve
    these things each week. You want to go into your next one-
    on-one and be able to show progress in these areas. When
    you can demonstrate to your boss that you are improving
    in the areas they are asking you to, you're showing that you
    listen to their feedback and are making a meaningful effort
    to improve yourself.

- Ask for advice on specific topics.

  - Your manager has been around much longer than you have.
    They've navigated difficult situations and have a wealth of
    knowledge and experience. Use that to your advantage and
    ask them how to deal with specific scenarios.

- ○ "How do I get better at saying no to requests that come in from other teams?"
- ○ "In your experience, what's the best way to deal with a difficult teammate who doesn't listen to my suggestions?"

Just doing these few things will help you get more out of your one-on-one meetings with your manager and provide you with plenty of concrete things for you to work on in order to grow as a software engineer. As long as you remember that your one-on-one time is meant to discuss personal and career growth opportunities and not status updates, you'll be able to make the most of the personal time you have with your manager.

The more you can demonstrate to them that you listen to their feedback and apply it in your day-to-day work, the more you will show them that they can trust you and that you deserve their respect.

> #### 𝒫 RESOURCES
>
> - The Simple Secret to Effective One-on-Ones[30] (effectiveengineer.com)
> - Why All Engineers Must Understand Management: The View from Both Ladders[31] (hackernoon.com)

6　How to Recover from Mistakes

At some point in every programmer's career, they'll go through the inevitable "oh crap" moment when some code they wrote suddenly breaks in production. It happens to the best software engineers, and it'll happen to you. No matter how many steps you take to mitigate risks, sometimes bad code will slip through the cracks and cause a catastrophic failure in production.

In some ways, this is a rite of passage on your journey to becoming an experienced software engineer, because you'll gain valuable experience

---

30. https://www.effectiveengineer.com/blog/secret-to-effective-one-on-ones
31. https://hackernoon.com/
    why-all-engineers-must-understand-management-the-view-from-both-ladders-cc749ae149
    05

identifying, triaging, fixing, and recovering from an incident in a high-pressure situation.

Software engineers work on complex systems. It's impossible to fully understand how each new code change you deploy will behave in a production environment. We can take measures to mitigate risks, but it's impossible to avoid them. So, what can you do if you're not able to completely steer clear of breaking code?

Good programmers accept that mistakes will happen. They don't know *when* one will happen, but they know what to do when things do go wrong. The ability to stay calm and collected and work through the problem under pressure, especially when alerts are going off and logs are filling up with errors, is a sign of an experienced programmer. They move with urgency and without losing their composure, because their main priority is getting the issue fixed and dealing with the impact.

Experienced engineers don't worry about their reputation or what their coworkers will think of them during a production outage. They know they may have to deal with some fallout after the dust has settled, but that's not their main concern when systems are down. While it's natural to be concerned about your reputation when things go wrong, that may hinder your ability to think through the problem clearly.

In How to Manage Risk,[10] we'll dive deeper into different things you can do to prevent future incidents. Here, we'll focus on things you can do *during* an ongoing incident. Let's look at different ways things can go wrong.

6.1 *Why Code Breaks*

We can't predict every dependency between our systems, or even all the dependencies between pieces of our logic in the same system. This alone makes it difficult to avoid introducing breaking changes, but it's not the only thing that contributes to broken code.

> ⟫ EXAMPLE

Let's look at other ways the code you write may break.

- **Untested code.** You may think it's a small change and you don't need to test it, or you may be in a hurry to fix a bug, so you put your code up for review as soon as you finish writing it.

While you may think you're working quickly, this is an easy way to introduce broken code into production because you didn't take the time to actually test it. While your code may look correct at first glance, it's possible your logic may have unintended behavior that you would never know about unless you actually ran it.

- **Unknown edge cases.** The data you use during development and testing may be clean, structured, and made up of expected values, but production data is often messy and varies greatly. Your system will need to handle inputs and events from your users (or other systems) that you didn't know about or account for when writing your code.

- **Missing context.** Oftentimes, the person who wrote the original code won't be the person who is updating or fixing it. Perhaps the original author is out of the office on vacation, moved to a different team, or moved on to a new company. When this happens, you may not have the full context about how part of the system works when you need to make modifications to it. There may be a specific reason the logic was written a certain way, or the logic may account for an edge case that isn't apparent when first reading the code.

- **Hidden dependencies.** As codebases grow, so does the dependency graph. You may deploy some code to production that you thought you had tested thoroughly, only to find that there is an obscure part of the codebase that also relies on the logic that was changed. Suddenly, you have another team asking you why their code, which they haven't touched in months, is throwing errors in production. Even worse, there could be bugs in a third-party or open-source library that your codebase relies on. These are often difficult to track down and can be difficult to fix if the maintainers aren't responsive.

- **Code rot.** Also referred to as software rot, code rot describes how behavior or usability of a codebase degrades over time, sometimes even if the code itself has not been modified. The environment the code runs in will change over time, or your customer's usage patterns may shift. They may request new features that build on top of existing logic, which can introduce new bugs. Even routine software maintenance contributes to code rot. You may need to update third-party libraries to patch security vulnerabilities, only to find the newer version breaks existing logic.

- **Environment differences.** As much as we'd like to keep our development environments in sync with our staging and production environments, it's extremely difficult to match them exactly. Your local environment variables may differ from what's running on production, which could lead to code that works locally but breaks on production. Differences in scale between environments can cause entirely new errors to occur in production systems that are hard to reproduce in smaller environments. Operating systems and their packages in higher environments may differ from your development environment. And finally, the hardware itself: while you develop your code on a personal computer, your code runs in production on powerful servers with different CPU and memory characteristics.

- **Third-party dependencies.** You may have external dependencies that your product relies on, such as third-party APIs that your code calls out to. When those services have outages of their own, they may affect your system and cause errors on your end. Even cloud hosting providers such as Amazon Web Services or Microsoft Azure go down from time to time and could bring your system down with it.

⚭ RESOURCES

- 9 Reasons Why Code Breaks[32] (git-tower.com)

6.2 *Murphy's Law*

The above examples are just a few ways in which your code can break—the list keeps going. Murphy's Law states that "anything that can go wrong, will go wrong." As a programmer, it's your job to identify all the scenarios in which your program can fail, and then to take steps to reduce the likelihood of those scenarios happening. In some cases though, your program

---

32. https://www.git-tower.com/blog/reasons-why-code-breaks/

will fail in unexpected ways that you never could have imagined, which makes it hard to plan for.

Here are more examples of ways that your system can fail. Remember, sometimes it's not just the code itself but other pieces of the system that can fail too.

- Bad logic
- Unanticipated inputs
- Bad configurations
- CPU maxed out
- Memory leaks
- No remaining disk space
- Hardware failures
- Network failures
- Database corruption
- User error
- DDOS attacks
- XSS attacks
- SQL injection
- Social engineering attacks
- Natural disasters

Any combination of these failures can occur at any given time, and you or your team might be responsible for getting the system back online when it does. The risk of some of these disasters can be managed and mitigated ahead of time, which we'll learn about in How to Manage Risk,[§10] but others are much harder to predict or prevent. The best thing you can do is be prepared for anything to happen at any time.

Now, let's shift our focus to things you should do during an incident that will help you triage, identify, and fix issues as they occur.

🔗 RESOURCES

- Murphy's Law[33] (wikipedia.org)

---

33. https://en.wikipedia.org/wiki/Murphy%27s_law

6.3 *What to Do during an Incident*

So, your code was deployed to production, and now things are getting thrown left and right. What do you do? It can be stressful, especially for someone that doesn't have a lot of experience dealing with production outages. The errors keep coming, and you haven't been able to identify the root cause yet. You don't even know where to begin looking.

First off, take a breath. Panicking won't do you any good here and will probably make the situation worse. So, the first thing you should focus on is staying as calm and collected as you can. As you get more experience working during production incidents, this gets more natural, but it's easier said than done the first few times.

Next, try to be methodical with your approach to identifying the root cause of the issue. There might be a million thoughts racing through your head about what it could be, but if you don't slow down, you won't be able to think clearly. Slow is smooth, smooth is fast. It may sound counterintuitive, but by focusing your attention on one thing at a time, you can often move quicker than trying to do too many things at once.

Try to eliminate potential causes one-by-one. And keep notes as to what changes were tried and which possible causes were eliminated, including why. That will help avoid rework and help document the final resolution, which may come in handy during a postmortem.

So, where do you start?

6.3.1 DETERMINE THE SEVERITY

There are whole books written on incident management, but almost every single one of them will mention some form of how to identify the severity of an incident. When you first hear that there is an issue in a production environment, one of the first things you should do is determine just how bad the issue is. In essence, determining the severity of an incident is the process of quantifying the impact that an incident has on the business.

Severity is often measured on a numerical scale, with lower numbers representing a greater impact than higher numbers. For example, a severity of 1 is an all-hands-on-deck type of incident, whereas a severity of 5 may be a minor, low-impact incident that can be fixed later.

Different businesses may use different severity scales and may even label them differently. One business may define a scale from Sev0 (high

severity) to Sev4 (low severity), while another business may define their
scale as Sev1 (high severity) to Sev3 (low severity).

Severity scores are useful for communicating the urgency in the midst
of an ongoing incident, and they also help communicate to business stakeholders what kind of fallout they can expect from a given incident. This
helps customer success and public relations teams when communicating
with external customers and partners, and helps the executive leadership
team understand the impact to the company.

Additionally, having a defined severity scoring system contributes to a
framework for who should respond to incidents and when others need to
be pulled into active incidents. The more impactful an incident, the more
important it becomes to have a plan in place for who should respond and
when they need to be notified.

≫ EXAMPLE

Here's an example of a severity framework with descriptions of
what each level entails:

SEVERITY	DESCRIPTION	EXAMPLES
Sev1	A critical incident with significant impact	A customer-facing service or feature (such as a checkout page) is down for all customers. Customer data is breached or compromised. Customer data is lost.
Sev2	A major incident with a large impact	A customer-facing service or feature (such as a push notification system) is unavailable for a subset of customers. Core functionality such as invoice creation is significantly impacted.
Sev3	A minor incident with a low impact	A bug is causing a minor inconvenience to customers, but a workaround is available. Performance for a page or feature is degraded.

Table: Sample severity framework.

These severity levels also determine the amount of post-resolution fol-
low-up, if any, is expected in order to prevent the same issue from happening again. In some cases, the most critical severity incidents require
communication with customers and other parties who were impacted.

The more clearly defined your severity levels are, the more likely your
team will know how to react and respond to an ongoing incident. If your
company already has clearly defined severity levels, it's a good idea to read

them and understand when each one should be used. This will help you know what to do and how you should respond during an incident.

Some companies have dedicated incident management or crisis management teams. In those cases, they will be responsible for defining the severity levels, determining the severity of each incident, and even deciding whether an issue should be classified as an incident at all or just a bug. These teams often get involved in higher severity incidents to help facilitate which teams need to be involved in resolving the incident. They also handle communication between teams as well as status updates to company executives and external customers, vendors, or partners. Learning how incident management teams operate is key to working well with them during high-stress incidents and getting issues resolved quickly. If you're able to hop into an incident and provide valuable and timely information to help get things resolved, you'll be able to build your reputation as a problem solver and leader within your company.

⸜ RESOURCES

- Understanding incident severity levels[34] (atlassian.com)

6.3.2 **CLUES, ANOMALIES, AND PATTERNS**

Look for clues, anomalies, and patterns. You'll want to observe and absorb as much information as possible and then figure out how it's all related, if at all. Hopefully, you have some internal dashboard, monitoring tools, or logs that you can dig into to find the information you need.

◇ IMPORTANT Look for these three things to help you figure out what's happening.

- **Clues.** Your goal here is to gather evidence.
  - When did the errors first start?
  - What changed around that time?
  - Did we recently deploy new code?
  - If not, when was the last time we deployed, and what changes went out?
  - Do we have any cron jobs that run around that time?

---

34. https://www.atlassian.com/incident-management/kpis/severity-levels

- Check your logs and any other observability tools.

  - Look for HTTP status codes.
  - Look for stack traces.
  - Read the actual error messages, and then read them again. What is it actually saying? Sometimes you might need to read a message a few times before you truly understand it.
  - Are errors coming from specific servers or all? What about containers?

- Check server health metrics.

  - What does the CPU utilization look like?
  - How much disk space is left?
  - How much free memory is remaining?

- Are bug reports coming from customers?

  - What were they doing when they experienced the error?
  - What browser are they using?
  - What operating system?
  - What steps did they take to trigger the error?

- **Anomalies.** Your goal here is to identify outliers.

  - Try to determine a baseline for your system under normal conditions.

    - What did the server metrics look like before the incident vs. now?
    - Are you seeing an increase in any metrics vs. the baseline, such as mobile devices hitting your API?
    - Are you seeing a decrease in any metrics vs. the baseline, such as a slowdown in processing your background jobs?

- **Patterns.** Your goal here is to find any repeating occurrences that could offer additional insights.

  - Try to find patterns in the errors across time.

    - Do the errors happen at certain times of the day? Do they line up with cron jobs?

- Do the errors happen at specific time intervals?

  - For example, the errors seem to be happening once every minute, or the errors seem to happen at random, with no pattern.

- Do the errors happen at specific times during the day?

  - For example, the errors seem to occur towards the beginning of the workday, right when users log on in the mornings.

- Try to find patterns in the data inputs when errors occur.

  - Do the errors happen when a user enters a number with an integer? Or just when they enter a number with a decimal point?
  - Are the errors happening with all input values, or just some? Maybe your logic doesn't account for an edge case when they select a specific value from a dropdown list.

In some cases, asking questions like the above may help you uncover the root cause, but not always. If you're able to ask the right questions, and then dig deeper to uncover the answers to those questions, you should be able to at least narrow down the problem.

6.3.3 OBSERVE HOW YOUR COWORKERS WORK

As you gain more experience and help diagnose and fix issues in production, you'll gain a better feel for what questions to ask based on the nature of the errors you're seeing. A big part of this just comes from experience, and that experience comes from observing how your coworkers handle these kinds of situations.

The senior engineers on your team have likely dealt with many more production incidents than you have, and they will demonstrate that experience when dealing with an outage. Use this as a learning opportunity.

Watch your coworkers and observe what they do during an incident.

- What monitoring dashboards are they checking? This will show you what metrics they consider important.
- How urgently are they moving? Is this an all-hands-on-deck incident, or something that needs to be fixed but isn't a major issue?
- Who are they communicating with?

- What channels are they communicating across?

You can learn quite a bit just by observing the senior engineers and how they conduct themselves during an outage. Before you realize it, you'll start picking up the same techniques and processes they follow, and you'll be able to diagnose and put out production fires in no time.

6.3.4 COMMUNICATE

This is arguably the most important thing you should remember to do during an incident. Collaboration tends to suffer when communication breaks down, and collaboration is paramount during a high-priority incident.

When in doubt, don't be afraid to overcommunicate to your coworkers. Let them know what you find, especially when searching for clues, anomalies, and patterns. You may not be able to connect the dots yourself, but you might offer a clue that helps your coworker do so. Or perhaps a coworker will mention something that helps you connect the dots and identify the root cause. The important thing is to share as much information as you can, so that you can work together to solve the problem.

You're part of a team, and you're all working towards the same goal—fixing the issue. It doesn't matter who discovers the root cause. It's more important that the root cause is identified and a fix is applied so that you can get back to the project you were working on.

6.3.5 TAKE ACCOUNTABILITY

There will be times in your career when you'll be pulled into an incident for something that's not your fault, but it's still your job to fix the problem without assigning blame. There will also be times when you'll be dealing with an incident that was caused by a change you made, whether it's bad logic, a bad configuration change, or an inefficient SQL query. When these things happen, it's best to be honest and own up to your mistake.

This may be much harder than it sounds depending on the severity of the issue, but taking responsibility when you make a mistake is *always* the best thing to do in these situations. Avoiding ownership is the worst thing you can do, because it can hurt the credibility you've been working on building with your coworkers and your manager.

After all, there's a good chance your teammates already know you're the one who made the mistake because they reviewed your code and know it's related to the outage. You can make all the excuses in the world, point fingers, or go hide in a conference room, but it won't change the fact that an incident happened that was related to code you wrote. It's all documented right there in the version control system, and it's easy for your coworkers to search the commit history to find your name next to the code that broke.

Your coworkers may start asking questions about your changes, but do your best not to view it as a personal attack. They may just be trying to gather more information about how the code is supposed to work or the context around why you made the change.

◇ CAUTION Whatever you do, try not to get defensive if people start asking questions. Immediately trying to justify a mistake or possible mistake is not going to help solve the problem and will make others defensive or frustrated, too.

》 EXAMPLE

Some programmers do this by pointing out that their pull request passed the code review and was approved by their teammates. This is bad because they're trying to shift the blame on to someone else for not catching their mistakes. Even worse, some programmers may try to blame the QA team for not catching their mistakes. Ultimately, you are responsible for your own code. So, it's not fair to rely on others to catch your mistakes and then blame them when they don't.

》 EXAMPLE

Other programmers may blame some constraint they had to work around, such as a legacy part of the codebase that hasn't been updated in a while. Once again, you're responsible for the code you write. Sometimes you need to work around constraints that are out of your control. That's part of the job as a software engineer—to solve problems within a set of constraints. Blaming your mistakes on something that is out of your control doesn't look good. Instead, it's better to admit that you hadn't considered an edge case or that you didn't understand the legacy system as well as you thought you did.

When you blame your mistakes on someone or something out of your control, you're playing the victim. And when you play the victim, you're admitting that you were not in full control of the situation. You're admitting that you didn't fully think through the ramifications of your code or that you didn't test your code as thoroughly as you should have. Not only does playing the victim look bad, but it will strain your relationships with your coworkers, especially if you're attempting to pin blame on them.

By being open and honest about where you fell short, you're allowing yourself to accept that you're not perfect. In a way, this makes it easier to learn from your mistakes because you're letting your walls down. It'll be easier to learn how you can avoid making the same mistake again in the future, rather than feeling like you are being judged for not knowing something.

Making mistakes can be painful, but they are opportunities to grow *significantly* as a software engineer. There's a saying among sailors that "smooth seas never make a good sailor," because you'll never be great if you never experience adversity. The skills you learn during the rough times in your career will be hard fought, but I can guarantee you'll learn to never make those mistakes again, and you'll come out of the storm much smarter and more experienced.

⚙ RESOURCES

- A developer's guide to programatically overcome fear of failure[35] (pagerduty.com)
- Don't Be Afraid to Break Stuff[36] (blog.codinghorror.com)
- Lessons learned in incident management[37] (dropbox.tech)
- My Most Embarrassing Mistakes as a Programmer (so far)[38] (stackoverflow.blog)
- How to Avoid the Biggest Mistake You Can Make as a New Software Engineer[39] (effectiveengineer.com)

---

35. https://www.pagerduty.com/blog/engineers-guide-on-the-importance-of-failure/
36. https://blog.codinghorror.com/dont-be-afraid-to-break-stuff/
37. https://dropbox.tech/infrastructure/lessons-learned-in-incident-management
38. https://stackoverflow.blog/2019/10/29/
    my-most-embarrassing-mistakes-as-a-programmer-so-far/
39. https://www.effectiveengineer.com/blog/new-software-engineer-mistakes

7 How to Ask Better Questions

Asking questions is a natural instinct in humans. In fact, it's so ingrained in our DNA that we begin asking questions about anything and everything from the earliest days of our childhood. Kids are curious about how the world works, so they ask questions as a way to help them make sense of the things they see. They're full of endless questions, but there's one that is simple, yet so powerful at the same time: *why?*

- Why is the sky blue?
- Why do I have to eat my vegetables?
- Why do I have to go to bed right now?
- Why can fish breathe underwater but I can't?
- Why is my sister taller than me?

Kids ask these questions because they see something but don't understand why it is the way it is. Children's brains are still developing, and they don't yet understand why they have to do certain things or why they can't do other things. So, they do the only thing they know how to do, which is ask questions.

Despite the fact that most answers to those questions are way more complex than what a child can understand, the mere fact that they're asking "why" is such a profound step in their learning process.

Asking why indicates a certain level of curiosity for understanding. They don't just accept something for what it is, they want to know the reason behind it. When you ask why, you're taking an active approach in gathering knowledge about something. You want to get to the truth because when you understand the reason behind something, you can build upon that foundation, connect the dots between ideas, and expand your knowledge.

As a software engineer, it helps to be curious about your craft. Just like a child asks questions to understand something about the world around them, you should ask questions about the codebases, systems, and team processes you work with on a daily basis. There's so much knowledge to be learned, and sometimes asking the right questions is the only way that wisdom will be passed down from senior engineers to junior ones. Your teammates are some of the most valuable resources you'll have during the course of your career. Combined, they likely have decades of experience between them, so it's a good idea to tap into that knowledge if you can.

> EXAMPLE

A curious software engineer might ask:

- Why is our data processing pipeline a series of asynchronous jobs instead of one big calculation?
- Why did we choose NodeJS over Ruby or Python for our back-end?
- Why do we spend so much time grooming our ticket backlog when we could be building?
- How does our fraud detection system differentiate a legitimate purchase from a fraudulent one?
- Why did we decide to use the builder pattern here instead of calling the constructors directly?

Asking these kinds of questions not only helps you learn and understand how the systems around you operate, it also gives you context behind some of the decisions that were made in the past, even before you joined your team, which will help you understand why your team might do things a certain way.

You'll learn a lot just by asking questions, but keep in mind that there's a fine line between being curious and being annoying. Just as parents often get fed up with the never-ending questions about "why this" and "why that," your coworkers may get tired of having to answer all of your questions. Remember, asking someone a question may interrupt their train of thought, or they may need to stop what they're doing in order to explain the answer to you. It's possible to ask *too many* questions, so it's important to find the right balance of figuring something out on your own versus asking someone to explain something to you.

Here are a few examples:

To ensure you're not interrupting your coworkers too often, it's good to do your own research prior to asking your question.

As a junior engineer, it can be intimidating asking senior engineers a question, especially if you're just starting out in a new position. Half the battle is just figuring out *what* to ask, and that requires a good understanding of the problem you're trying to solve. Once you know what to ask, you'll also need to think about how to ask it and who you should ask. Let's look into the different types of questions you may find yourself asking.

🔗 RESOURCES

- Please ask stupid questions as a new software developer[40] (nikitakazakov.com)
- When Should I Interrupt Someone?[41] (zwischenzugs.com)
- 3 Ways you can ask Better Questions as a Junior Developer[42] (dev.to/erikmelone)

7.1 *Asking for Help*

There will be times when you'll hit a dead end with your code. Either you're getting an error or your code just isn't doing what you think it should be doing. It's tempting to drop everything and ask a coworker for help, but you shouldn't ask them without being prepared.

Knowing *what* to ask is often the hardest part. You're struggling to get the code to work because you don't have a good grasp on the problem you're trying to solve, but where do you even begin asking for help?

A simple thing to remember is that you should be asking for advice, not a solution. Don't expect your coworker to fix your bug or code your feature for you. While most people will be happy to help you arrive at a solution, no one wants to do your work for you.

When you ask someone for help, you're asking someone else to take time away from their own projects to help with yours. Switching context back and forth is hard when working on something that requires deep focus, and you need to respect your coworker's time if you want to build trust with your team. Try not to waste their time with a question that you can easily find by querying a search engine.

So how can you do that?

- Come prepared with a clear explanation of the problem you're trying to solve. If you can't explain the problem clearly, you're not ready to ask for help yet.
- Give examples for how to recreate the problem if you're able to.
- Show them what solutions you've already tried so they can eliminate potential options.

---

40. https://www.nikitakazakov.com/ask-stupid-questions-as-software-developer
41. https://zwischenzugs.com/2021/03/15/when-should-i-interrupt-someone/
42. https://dev.to/erikmelone/how-to-ask-the-right-questions-as-a-junior-developer-5701

- And remember: ask for advice, not a solution.

The more effort you put into asking the question, the less time your coworker will need to get up to speed with the problem. You'll help your coworker understand the problem quicker, which means they'll be able to help you quicker.

7.1.1 WHAT TO ASK

Don't be discouraged if you're having trouble explaining what the problem is that you're trying to solve. Oftentimes, it's a complicated explanation involving lots of context before you can even get to the problem itself. It may be hard to find the right words to explain what the problem is, so it's important to take the time to understand the problem completely before asking for help.

If you don't understand the problem well enough to explain to someone else, take some time to review everything again. Reread the description of the task to make sure you're understanding it correctly. In some cases, it helps to write the problem down on paper first or type it out in your notes. Getting the thoughts out of your head will help you organize them, and sometimes, the solution will come to you during this time.

Consider the following:

The second example conveys more information to the person helping you, and it gives them enough context to know what you're trying to do and where you need their help. Try to be specific in the details of your question, and the more context you can provide, the better.

7.1.2 NARROW DOWN THE PROBLEM

It sounds silly, but if you're getting an error, make sure to read the error message. I mean *really* read it. Read it multiple times to make sure you're fully understanding what the compiler or interpreter is telling you. Sometimes the answer is staring right at you from the error message, but you may not even realize it because you're skimming over it.

If you have a stack trace, follow the code path line by line. The compiler is giving you hints at what the issue is and where you can find it, so follow the trail. Pull up each file and go to the line listed in the stack trace. Do this for the last few functions that executed before you hit the error.

If there isn't a stack trace available, try modifying the code in your local environment to add some logging. Add log statements to print out the information you need to figure out what's happening.

Better yet, use a debugger if you have one set up. Being able to step through the code and see the values change feels like superpower when tracking down bugs. Not only that, but you'll be able to see how your logic works in real time, with real inputs. You'll come away with a better understanding of what the code is doing and how that part of the system works.

7.1.3 REPRODUCE THE ISSUE

Next, make sure you can reproduce the issue before asking anyone for help. If you can't reproduce the issue, then you'll just be wasting other people's time by asking for their help. You'll want to be able to show whoever is helping you how to reproduce the issue to give them context on what you're trying to solve. After all, if you're not able to get your application in the right state to trigger the error, how will you even know that you fixed it?

◇ IMPORTANT   There are exceptions to this rule, however. For example, if you've spent three days trying to reproduce an issue and you're still not able to, it might be good to ask for some help. Use your best judgment and only ask for help if you've put in significant effort to reproduce the issue and you're still not making any progress.

Try different inputs, both good and bad. Does the error only show up with specific input types, or does your program throw an error regardless of the kind of input. Check the same thing but with different application configurations as well. For example, does the export functionality break for all file types or just a subset of available file types your application provides?

7.1.4 GIVE CONTEXT

Give some background context about the problem to the person helping you out. Don't assume they know what problem you're trying to solve, and don't assume they're familiar with the part of the codebase you're asking them about. If possible, try to give them some context around the problem to help get them up to speed. Show them what solutions you've already tried, and explain to them where you're getting stuck. Are there solutions you've been able to eliminate? Give a quick explanation about how you were able to determine that certain solutions won't work, and why.

The more information, the better, so try to be as specific as possible. Does your data need to be in a certain state in order to reproduce the bug?

Let your coworker know. Are you able to reproduce the bug with all third-party integrations, or just a specific one? Let them know. Give as much context as you can, because that will help them narrow down the problem and eliminate potential scenarios.

7.1.5 DO YOUR RESEARCH

There's a misconception among young developers that experienced programmers know so much that they rarely need to Google anything, when in reality, it's just the opposite. All programmers use search engines and Stack Overflow to help answer their questions, even the senior and staff engineers.

Humans can only retain so much information before we begin to forget things. Our brains can hold vast amounts of knowledge, but there's still a limit to how much information we can retain.

Chances are you're not the first person to encounter an error message or run into an issue with your implementation. There's a good chance that there are plenty of people who have posted similar questions to online communities or written blog posts on the same issues you're running into.

Try searching Stack Overflow, reading blogs, or reading GitHub issues and pull requests. Those are the best places to start. Be descriptive in your search queries, but try to remove any specific file or variable names that are unique to your code. You may need to try a few different search queries before you find what you're looking for, but if you're able to explain the problem correctly with your query, you shouldn't have any trouble finding someone else who also ran into that same issue.

⟡ CAUTION   If you can find a solution to your problem by searching the internet, that's great. That means you didn't have to ask any of your teammates for help! But be careful though. While it's tempting to copy and paste solutions you found on the internet, there are many examples online that may not be the most efficient code or the most secure, and the last thing you want to do is compromise the security of your codebase using a solution you found on the internet. If you're unsure of something, ask a teammate to help you evaluate the quality of a solution first.

7.1.6 SEARCH INTERNAL DOCUMENTATION

It's easy to forget that your team's internal knowledge base is a good place to look for answers. Engineers often document how a part of the system works, including any known issues, limitations, or other considerations and trade-offs they made when designing a feature.

If you're running into issues setting up your environment, getting your build to compile, or running your test suite, the solution to your problem may be documented in the internal knowledge base. It doesn't hurt to take a quick look there first.

While you may find an answer in the internal documentation, that page may be out-of-date by the time you find it. If you find an answer, try to find the original author and ask them to verify that it's still up-to-date. If the original author is no longer on your team, ask a teammate who you think would be familiar with that part of the system.

There's a lot to consider when asking for help, but in some cases, you can avoid it altogether if you're able to leverage the tools around you.

Sometimes, you'll run into scenarios where you'll need to ask questions to gather more information before you can proceed. In these situations, you'll need to ask questions to *clarify* something.

7.2 *Asking for Clarification*

When you're working on a new feature, a bugfix, or almost any project, you may not have all the information you need in order to complete the task. The devil is in the details, and an experienced senior developer knows when they need to ask the project owners to clarify small but important details before proceeding.

Sometimes, the small details won't change the implementation much, but other times, a slight clarification to the project's requirements could have major ramifications for the design and implementation of your solution.

Here are a few things to consider when you may not have all the information you need to move forward with a project.

7.2.1 BREAK DOWN THE REQUIREMENTS

Before asking for clarification, first try to break down the requirements into the smallest pieces of individual features you can. Try to think about your task in terms of the following requirements:

Inputs. Determine which inputs are required, which ones are optional, and what the expected values or ranges should be for each input. This helps you dig deeper into what inputs you should expect so that you can build logic around them to prevent bad data from getting into your system.

Try to think in terms of how the system will handle different scenarios. You will naturally come up with some additional questions for your teammates or the product owner if you can think of edge cases where a user may enter some unexpected input.

> ❯ EXAMPLE

- The name field is only one input field in the design. What if someone only gives us their first name when we need their full name? Should we break it into two fields so we can make each one required?
- Are we verifying emails to make sure they are deliverable, or is basic email format validation good enough?
- Will the number always be an integer? Or should we expect decimals too?
- Will the data streaming from the widget's sensor always be in a range between 0.0 and 1.0? What should we do if we receive a negative number from the sensor? And can this value be null if there is an issue with the sensor?

The more clarification around your inputs, the more robust your validation logic and error handling will be. Common errors occur when a program encounters unexpected inputs, so if you can ask good questions to clarify any missing requirements, you'll be able to prevent errors before they happen.

Outputs. Double-check what type or format your outputs should be in, especially if you expect the output to become the input in another program or function. This is easier if you're using a strongly typed language, and it is critical if you're using a scripting language.

> ❯ EXAMPLE

Here are some examples of what to ask when clarifying output requirements:

- Should we output the data in CSV, TSV, JSON, or give the user options for all three?

- Does the data we're outputting contain user-submitted content? Does our templating framework escape outputs automatically? How can we avoid XSS attacks?
- How many results do you want in the API response? Should we paginate the results?
- What should the precision on our floats be when we format them during rendering?

Error handling. Reliable software programs contain robust error handling. Reliable programs gracefully recover from certain errors and display helpful information to the user if the program is unable to proceed. In order to write code like this, you have to anticipate ways in which your program can fail. If you understand all of the things that can go wrong with your code, you can then add logic to handle those errors gracefully or to exit the program and display a helpful error message to your users.

> ❯ EXAMPLE

Here are more example questions related to error handling:

- If the API request fails, how many times should we retry? Should we only retry for specific status codes?
- What error message should we display to the user if they hit [insert rare corner case here]?
- Should I throw an exception if we hit [insert weird edge case here]?

If you're unsure of how to proceed when a block of code may fail, ask your coworkers or the project stakeholder. They will often help clarify how you should handle errors in certain situations.

Business logic. The business logic encodes into your program the core set of business rules that determine how it should store, modify, and delete data. Many times, this logic will closely resemble real-life concepts such as inventories, invoices, accounts, or processes.

Making assumptions in business logic can lead to misunderstandings in how the business operates, so it's always important to clarify any ambiguity before proceeding. A small bug in your business logic can have enormous downstream effects such as data loss, duplicate data, or many other unintended consequences.

> ❯ EXAMPLE

Here are a few example questions for clarifying business logic.

- Should we start the free trial when the user creates their account or after they verify their email?
- Should we send an alert as soon as the sensor temperature reaches the high heat threshold, or after it's above the threshold for more than X seconds?
- Should we disable the form after the user submits their answers, or should we allow multiple submissions?

If you don't have all the information you need to complete a task, you'll need to ask for clarification. It's your responsibility to find gaps in the requirements and to ask questions that will clarify any ambiguities and fill in those gaps. The better you're able to fill in the missing requirements, the better software you'll write, and that starts with communicating clearly to others when you aren't sure how to handle specific scenarios in your code.

7.2.2 MAKING ASSUMPTIONS

Programmers enjoy autonomy in their role. When you're starting out in your career, you'll need some help when completing your tasks, and that's okay. By the time you start a new task, a lot of the big decisions will already have been made for you, and your job will be to implement a predetermined solution. This is good because it allows you to focus on writing quality code rather than trying to implement a solution where the end result isn't crystal clear.

When you are implementing a set of requirements someone else has already figured out, you can focus on good coding fundamentals and understanding how your changes fit into the larger context of the codebase. In a well-architected system, sometimes you'll be able to add powerful new functionality by making a few small tweaks, and in doing so, you'll see firsthand how good code should be written. You'll pick up new ideas over time and form your own opinions based on what you've seen work well in the past. In the process, you'll build confidence to make more decisions on your own in the future.

As you gain experience and grow into a seasoned developer, you'll naturally want to make more decisions on your own rather than be told how to implement solutions. This is a good thing, but it can be a difficult time in your career to navigate. At some point, you'll find yourself at a crossroads where you're confident enough in your ability to devise a working solution, but your ideas may not be sufficient for what the senior engineers had in mind.

You'll disagree on how some solutions should be implemented. These conflicts are difficult because emotions often get the best of people. Sometimes, you'll be right, but other times, the senior engineers will rely on their experience to override your decisions. Sometimes, they will have legitimate reasons based on experience to push back on your decisions, but other times, it may come down to personal preference.

It's easier said than done, but as a junior engineer you need to do your best to take your emotions out of the development process. If you can separate your decision-making process from how you view your coding ability, you'll be able to navigate through your career better.

As you gain experience, you'll encounter ambiguity in requirements that might seem insignificant in the context of the task, and it's natural to want to make decisions on your own. You were hired because you're smart, and your employer values your ability to make good decisions, so why wouldn't you want to make certain decisions on your own? After all, Steve Jobs famously said that "it doesn't make sense to hire smart people and tell them what to do; we hire smart people so they can tell us what to do."

The longer you work as a software engineer, the more you'll gain a better feeling for what to do when faced with ambiguity, but be careful in these situations. Making your own decisions can be rewarding, but if you're not careful, you may be making decisions based on flawed assumptions.

An assumption is a prediction that something is true without proof or evidence. The thing that makes assumptions dangerous is that they're often made based on experience, and as a junior engineer, you may not have enough experience to safely make certain assumptions. Senior engineers, on the other hand, may have relevant experience to contradict your assumptions, which often leads to conflict.

When a senior engineer pushes back on your solution, you may feel like you're being personally attacked, but try to view their experience as an opportunity to learn *why* they're pushing back. Most of the time, it's as simple as asking why they think your assumption is wrong. You may be surprised that they mention some reason you hadn't considered or some edge case that you didn't think was possible.

The important thing to understand in these situations is that there are still a lot of things you don't understand about software development. The same can be said about senior engineers as well, but it's especially true for those who have only been working professionally for a few years. You may think you know quite a bit, but the reality is that you've only scratched the surface.

So how do you avoid these conflicts while you're building up the confidence to make your own decisions? The answer is to make assumptions, but verify those assumptions before implementing any of your decisions. It's a subtle detail, but it's an important one. You can still make decisions on your own when the opportunity arises, but verify with your manager, a senior member on your team, or a project stakeholder before proceeding with any development work.

When you verify your assumptions before beginning any coding, you'll avoid any situations where you'll have to redo your work because you made a decision that wasn't correct or that someone disagreed with. It's faster and cheaper to refactor an idea than it is to refactor code, so it's always good to double-check with others that your assumptions are correct and that you're not missing some important context before implementing your decisions. If your assumptions are correct, that's great! You're getting smarter and your experience helped you come to the correct decision. If not, ask for clarification on why your assumptions were incorrect, or why your decisions won't work in the given situation. And by coming to a conclusion before writing any code, you'll be able to avoid conflicts after you've put in time and effort coding an incorrect solution. It's a win-win for everyone involved, and in the worst case, you'll walk away learning something new about why your decision was flawed and what you can do better next time.

7.3 *Asking to Learn*

At the beginning of this section, we discussed why children ask questions, and how it helps them connect ideas and form a better understanding of the world around them. In a way, most engineers also do this without even thinking about it. By asking the different types of questions you've learned about in this section—asking for help and asking for clarification—you'll naturally learn things from the answers you're given. You'll learn why you should choose one software design over another, or why you should consider a certain edge case when you didn't think it mattered.

In some cases though, you'll just have curiosity for how something works or why something is the way it is. You won't necessarily be asking for help or to clarify anything, but you'll still have questions. In this case, you're just asking a question for the sake of learning.

Sometimes it's good to challenge the status quo and ask your teammates why something is the way that it is. In some cases, the answer may be underwhelming. You may receive an answer such as "well, that's just how we've always done it." In asking these questions, you're challenging the other engineers to think through why they've always done it that way, and what they could do to improve a process or part of the codebase.

In other cases, there may be a very good reason for why something is done a certain way that you may not have realized. There may be historical context that you were unaware of because you joined the team after some important decisions were made. Perhaps you never would have known this had you not asked the question in the first place.

In some cases, asking questions will help you understand your team's processes, which will help you be a better contributor in the future. Other times, asking questions will help you understand why a part of the system is architected a certain way, which will help you write better software in the future. Or there may be times where asking questions helps you understand someone else's thought process and how they approach solving a problem. It can be eye-opening to see what considerations and trade-offs someone else takes into account when making decisions, and this can often influence your own thought processes in the future. Asking good questions helps to uncover new ideas, techniques, and processes that you can leverage to become a better software engineer.

7.3.1 WRITE IT DOWN

When you read about a new idea for the first time or hear an explanation for why some status quo is the way that it is, it can be revealing. In some cases, it can change your perspective on a topic or how you view certain things. Over time, these new ideas may fade out of your memory, and you'll soon forget about them.

It's always helpful to have a good note-taking system in place to keep track of small task lists and meeting notes, and a good habit to get into is to document new ideas and concepts as you learn them. This can be in a personal note-taking app or your private space on your company's knowledge platform.

First, writing things down helps you internalize new concepts and retain new ideas by forcing your brain to process the information in a more detailed way. It's often said that the best way to learn something is to teach someone else, so try writing your notes in a way that teaches your future self what you want to remember. Not only will it help you internalize what you learn, but you'll be able to review and reference your notes in the future.

7.3.2 NEVER STOP LEARNING

Software engineers are curious creatures. We strive for continuous improvement and are constantly looking to refine our understanding about how the world works around us. We apply years' worth of experience to the code we write, but more importantly, we never stop learning from the decades' worth of experience that our colleagues bring with them.

Asking questions is a fundamental aspect of our jobs as software engineers, and asking good questions is equally an art and a science. Over time, you'll hone your ability to ask better and better questions, but it's always important to keep in mind which kind of question you're asking and the context in which you're asking it. Sometimes you'll get the answers you were looking for, sometimes you'll get unexpected answers that change your perceptions, and sometimes you'll get insufficient answers that will require additional questions. If there's one thing for certain though, it's that you should never stop asking questions, because asking *why* is fundamental to growing as a software engineer.

🔗 RESOURCES

- How to ask good questions[43] (jvns.ca)
- How to get useful answers to your questions[44] (jvns.ca)
- Asking Questions[45] (zalberico.com)

---

43. https://jvns.ca/blog/good-questions/
44. https://jvns.ca/blog/2021/10/21/how-to-get-useful-answers-to-your-questions/
45. https://zalberico.com/essay/2017/02/21/asking-questions.html

- Good Programmer vs Average Programmer - and, Why Asking questions and Paying attention to Details matters[46] (dev.to/javinpaul)

8 How to Read Unfamiliar Code

It's a common misconception among students and aspiring programmers that professional software engineers spend all of their time writing new code and building new systems from scratch. Many new developers face a rude awakening when they land their first job and find out that this is far from the truth. In fact, aside from planning and documenting, most of your early-career time will be spent maintaining, extending, and fixing bugs in legacy codebases. You'll be tasked with making small- to medium-sized changes to the code that your team members wrote, and you may sometimes find yourself working on code written by someone who is no longer with your company.

Working on legacy code gives you the opportunity to get experience working on a mature codebase. In a way, it can be seen as a rite of passage on some teams because it allows you to get familiar with complex abstractions and business logic. There will be design patterns, coding standards, and test cases that the previous programmers established and that you'll be able to follow when making your changes. Following established patterns when learning a new codebase will help you focus on the behavior of your code without getting too bogged down in details about the design and architecture of the code.

This is especially true when you join a new team, because you'll be learning the nuances of the codebase and the business rules while getting up to speed. Your manager will probably start you off with some small bug fixes and enhancements before you graduate to larger projects. In many cases, it would actually be counterproductive for you to jump in and make large changes to a codebase that you don't understand very well. That would be very risky, especially as a junior software engineer still learning the best practices.

---

46. https://dev.to/javinpaul/
   good-programmers-vs-average-programmer-and-why-asking-questions-and-paying-attenti
   on-to-details-matters-j3h

Before you can run, you need to learn how to walk, which is why it's so important to develop skills for reading and understanding unfamiliar code. The quicker you can read code and understand its intended behavior, the quicker you'll be able to make changes, fix bugs, or identify edge cases that weren't considered.

Your manager will give you projects that will require you to do some digging to identify the location of a bug or to determine the best way to extend a feature to enhance its functionality. At times, you'll feel like an archeologist uncovering corners of the codebase that haven't been touched in years, decoding what the previous engineers were thinking when they wrote the code on your screen, and piecing together a mental model of how the system works as a whole.

Even though you may read a piece of code and understand its behavior, you may not have all the information you need in order to fix certain bugs. Code can be very nuanced sometimes. You may read a piece of code and think there's a better way it could have been written, or that perhaps the problem could have been solved in fewer lines of code, but there may be additional context that the original author had to consider but that you may not yet understand. Your job is to put yourself in their shoes and figure out what their code is doing and if there's a reason why it was written the way it was. Oftentimes, the author had to accommodate specific edge cases that may not be apparent on a first reading of the code. You'll need to put on your investigator hat and ask yourself some questions about their code.

> ❯❯ EXAMPLE

Here are examples of things you'll need to figure out as you read new code line by line:

- What kind of inputs did the author expect? Are they validated?
- Which edge cases did the author consider? Are there any that aren't handled?
- What do the data structures look like?
- What assumptions did the author make about the data? Could any of those assumptions be wrong?
- How did the code change over time? Were additional changes made after the code was shipped?

Reading other people's code isn't the most glamorous aspect of being a software engineer, but it's an important skill to master if you want to excel in your career. It's frustrating reading code that's hard to follow, especially when there are layers of abstractions or it's written differently from how you would have approached the problem.

Reading other people's code might not have been what you had in mind when you decided to be a professional programmer, but it's part of the job. What might surprise you though is that reading unfamiliar code is also one of the most important things you'll do in your programming career. In fact, reading other people's code is one of the best things you can do to improve your own coding skills.

As a child, you were assigned to read books and write essays, which was no different. You had to read another person's work and come to conclusions about what the author's intentions were. Except in that case, you were dealing with literature and written language instead of computer programs and code.

Even the most successful authors in history didn't create their work in a vacuum. Ask any famous author what their favorite books are, and you'll receive title after title of books that inspired their own writing. In fact, some of the best writers often spend more time reading other author's works than writing their own. And they're not just skimming through the books, they're *studying* them: dissecting and analyzing the choice of words, sentence structure, style, tone, and vivid scenery. They notice which literary rules the original author followed and, perhaps more importantly, which ones they broke. By observing how great authors bend the rules of their language, writers become better at their craft, and they adopt similar techniques and styles in their own writing.

The same is true for software engineers. You must study other programmers' code in order to understand how their programs work. You'll learn new design patterns, ways to structure your codebase, optimization techniques, algorithms, novel solutions to complex problems, and so much more.

Reading code from better programmers will help you become a better programmer, plain and simple.

You'll mostly be reading code written by your coworkers, which is great because you can ask them about specific details when you have questions about their code. You may be reviewing their code in a pull request or reading code in a specific part of the codebase you're working on. Your team members are an excellent resource for learning, so make sure you utilize them when you're having trouble understanding a specific piece of code. Don't hesitate to ask questions if you don't understand a piece of code.

Additionally, with the rise of open-source software, you have an incredible amount of resources available to you online. Reading code from popular open-source projects is an excellent way to learn how other programs are structured, and you can follow along in the open issues, pull requests, and discussions around how new features and bugs are fixed and merged into the main branch. GitHub, GitLab, Bitbucket, and other websites have millions of open-source code repositories available online, so it's easy to find some popular projects in your favorite language. You can even subscribe to get updated on all new issues if you find a project you want to follow along with.

So, now that we've gone over the benefits of reading code and *why* you should read other people's code, let's jump into some specific tools and techniques you can use to improve your code-reading skills.

8.1 *Find the Entry Points*

First things first—figure out where the program starts. To execute a program, the loader (typically an operating system) will pass control of the process to a program's entry point, which begins the run-time execution of the application.

The entry point is the place where a program begins, and it's important to know what the program is doing once it begins executing the code. When you follow a program from the entry point, you'll be able to follow the application as it boots up and configures itself to do whatever work it was designed to do.

Some programming languages may enforce conventions for how or where a program should start, while others may give more freedom in how a program is executed.

- C-family languages, such as C, C++, and Rust, and JVM languages such as Java contain a predefined function called main.
- Interpreted languages like JavaScript, Python, Ruby, and PHP will simply begin execution at the first statement.

Once your program has control of the process and has begun execution, it will be able to access command-line arguments and environment variables that can be used to dynamically configure the behavior of your application during run time. The program may contain specific logic to check for these arguments or environment variables in order to change the run-time behavior of the application without needing to recompile or redeploy the application.

It's important to know where and how your program starts because that may give you valuable information as to how the program is configured, which could affect how the program behaves. If you don't know what run-time configurations your program is using, you may not fully understand what it's doing, so this is always a good first step.

8.2 *Leverage Your IDE*

Your integrated development environment (IDE) is one of the most important tools you will use when reading code. Your IDE gives you a set of tools to analyze and manipulate your codebase, so choosing a good IDE will help you navigate the code efficiently.

When reading code, you'll want an IDE that lets you jump to a function definition. This feature is crucial for learning and studying a new codebase, and most modern development environments should support this functionality. This allows you to jump through the codebase to see where a function is defined, which is useful whenever you come across a function call you're not familiar with.

This feature gives you the ability to step through the codebase and follow the execution path, which helps you build a mental model of the code and what it's doing. It's a great way to explore unfamiliar code and can help you get up to speed quickly.

When you jump to a function, take note of the file name and directory structure where the function lives. You can learn a lot about the structure of an application just by observing how things are organized.

Most IDEs that allow you to jump to function definitions should also give you the ability to move in the opposite direction as well. When you're looking at a function, you might want to know all the places where it's used within the codebase, which is helpful if you're trying to track down a bug or refactor a piece of code. The ability to see all places where a function is called is equally as powerful for learning and understanding a code-base.

If your IDE doesn't offer these basic features, consider switching to one that does. Once you get in the habit of navigating around the codebase by jumping from function to function, you'll wonder how you ever lived without it.

8.3 *Dig Deeper*

Development tools aren't perfect, and sometimes our IDEs won't be aware of the entire structure of the codebase. Perhaps you have some code that is called dynamically or your language supports metaprogramming, both of which can be difficult for IDEs to understand. In some cases, you may need to use other tools like grep or git grep instead, which give you the ability to search your codebase for specific patterns such as variables, functions, class names, or constants.

For example, you may come across a function called findNearbyLocations() while reading some code. In order to find all locations where that function is called, you can run the following command from your projects root directory:

```
$ grep -r findNearbyLocations *
```

That command will recursively search all directories in your codebase and output the lines where the term "findNearbyLocations" occurs. With this information you can pull up each file to see how that function is used. When you can see where a certain term is used throughout the codebase, you gain a better understanding of what the program is doing.

Most of the time, you'll want to search recursively using the -r flag, although this means it will also search in folders we may not want to query, such as dependency directories that contain large amounts of third-party code. While grep gives you the ability to exclude certain directories from your search, it may be annoying to have to manually exclude them every time.

Fortunately, if you're using git for version control, there is a command called git grep that works similarly, except that it automatically ignores any files and directories that are defined in a file called .gitignore. This makes it much easier to query your codebase without having to sift through files and directories you're not interested in.

With these tools, you have a way to query your codebase any time you come across a function you're not familiar with. This will help you learn how a function works, what parameters it expects, what the return values are, and where else it's used in the codebase. Using these tools will help you to better understand what the code is doing and how it is organized, and will ultimately help you build and refine your mental model of the codebase.

> 🔗 **RESOURCES**
>
> - grep — Reference Page[47] (man7.org)
> - git-grep — Reference Page[48] (git-scm.com)
> - ripgrep — GitHub Repository[49] (github.com)

8.4 *The Blame Game*

When you're reading through code, you may want to know when it was last changed. If you're using `git`, there's another tool called `git-blame`, which displays the last revision and the author who most recently modified each line of a file that you're interested in. This is useful for determining when certain functions were last modified and by whom.

Use the command below to view the last revision and last person to touch each line of a file:

```
$ git blame <file>
```

🔥 CONFUSION It should be mentioned that `git-blame`'s intentions are not to actually *blame* someone for writing a bad piece of code, and hopefully you won't use it for that purpose. It's simply another tool at your disposal for understanding the code and how it evolved.

You should consider using `git-blame` when working on a bug you've been assigned to, or when you have questions about a specific function.

---

47. https://man7.org/linux/man-pages/man1/grep.1.html
48. https://git-scm.com/docs/git-grep
49. https://github.com/BurntSushi/ripgrep

Git-blame will give you clues as to who you should talk to first when you have a question regarding specific lines of code.

Depending on the age of the codebase, the most recent author may no longer be with your company. If that's the case, you won't be able to ask them any questions, but you're not out of luck. With `git-blame`, you will still be able to find the commit hash, which you can use to view the full context of the changes. Oftentimes, being able to read the commit message and see all the other changes that were made in the same commit will give you more context for why the change was made.

If you're still not able to find any developers who are familiar with the code you're looking at, use `git-blame` to find the developers who made modifications to other parts of the file and ask them if they're familiar with the code in question. Chances are you'll be able to find someone who has worked in that part of the codebase before or reviewed the pull requests for the code in question.

> 🔗 **RESOURCES**

- git-blame — Reference Page[50] (git-scm.com)

8.5 *Read the History*

While `git-blame` shows you who made the most recent changes to each line in a file, sometimes you might be more interested in the history of a single file and how it's changed over time. Git offers a useful tool called `git-log` that lets you inspect the commit logs for a given file.

Use the following command to view a reverse chronological list of commits where changes were made to a file:

```
$ git log <file_path>
```

This will give you a full history of all commits to the file so you'll be able to see who made changes to it and, more importantly, *when* they made those changes. Just as with `git-blame`, you can use `git-log` to find the developers who made the most recent changes to a file, because they should be the ones you reach out to first.

---

50. https://git-scm.com/docs/git-blame

If you suspect a bug is located in a certain file, use git-log to view when a file was changed and by whom. It's extremely helpful if you know when a bug was first reported or when an error started popping up in your logs. You can use git-log to line up errors with changes made to specific files, which may help you pinpoint when bugs may have been introduced into the codebase.

    ∂ RESOURCES

- git-log — Reference Page[51] (git-scm.com)

8.6 *Log Some Data*

As you're reading through code, you will need to hold a mental model of the data in the system and how it is manipulated as the business logic is applied. Some code may be easy to follow, but you may find yourself deep in the codebase without any idea what the data looks like when it reaches a certain function. In these situations, it's sometimes useful to lean on your logging system to print some data to your log files so that you can inspect it.

Add a few log statements with data you're interested in. This could be certain values of variables or object properties, or it could be an arbitrary text string that will give you some useful information if you see it in your logs. Either way, setting log statements throughout your code is a quick and easy way to get a snapshot of what your data structures look like at a point in time when the code is executing. Sometimes, a well-placed log statement can reveal a bug you've been tracking down, or it can expose certain things that help you understand what the code is doing.

All programmers rely on logging to gain insight into what their code is doing, so don't feel like it's the wrong way to debug your code. Even the most experienced engineers rely on logging when they're developing new features or tracking down a hard-to-find bug.

---

51. https://git-scm.com/docs/git-log

8.7 *Fire Up Your Debugger*

Occasionally, you'll come across code that you won't understand no matter how many log statements you add. Wrapping your head around confusing code is frustrating, especially if you're trying to figure out how some piece of data is being manipulated. While you may be able to figure it out with enough log statements, it's messy to add them all over your codebase just to piece together what's going on. Sometimes a debugger is the better tool for the job.

When you distill a program down to the simplest form, it's really just taking some inputs, manipulating the data structures, and producing output somewhere. To really get a grasp on how everything works, you need to understand how the data changes as it moves through the system. While it's helpful to read through code and build a mental model of what the data structure looks like, it's sometimes easier to visualize the program with a debugger and observe how the data changes as it moves through the system.

If you have a debugger configured, you'll be able to see what the data looks like at each breakpoint you set. As you step through the debugger, focus on the data and how it changes as you step in and out of functions.

> 🔗 RESOURCES
>
> - Improve your debugging strategies[52] (functionize.com)

8.8 *Tests Contain Context*

An underrated technique for studying an unfamiliar codebase is to read through the automated tests. While it's not the most glamorous part of the codebase, there's an enormous amount of institutional knowledge stored in the test files. Automated tests are where past and present developers have codified the specifications the application is expected to operate within.

Most young developers don't realize that a mature test suite will show you exactly how a program should perform, because each test that's added to the suite should be designed to validate a specific part of the program

---

52. https://www.functionize.com/blog/improve-your-debugging-strategies

for a specific scenario. As you read through the test cases, you'll see what edge cases the tests handle and what the expected outcomes should be.

Additionally, the assertions in automated tests will show you what the expected output should be when you call a function. Assuming the tests are passing, this gives you a clear picture of how the system works and what application states you should expect.

8.9 *Don't Try to Understand It All*

Codebases are complex, plain and simple. A codebase's complexity can be roughly estimated as proportional to the number of engineers who have contributed to the codebase multiplied by its age. As more developers contribute to a codebase over time, the complexity continues to increase.

It's almost impossible to understand every line of a codebase, especially if you didn't write it yourself. In fact, even a solo developer who has written every single line of a codebase will forget the details and context of parts of the system over time. They may come back to a file they wrote months ago and struggle to remember how it works.

Setting the right expectations now will help reduce your frustrations in the future. It's *okay* if you don't understand how every line of code in a program works.

As developers, it's our job to form a mental model of how a program works, and how the pieces fit together to form a complete system. You have a limited capacity in your brain to hold this mental model, and eventually, you'll hit a saturation point where you're not able to hold the entire mental model in your head at once. As you learn new parts of the system, you may forget other parts you haven't visited in a while. It's natural and common among all software engineers.

Depending on the size of the codebase, it may even take years to feel like you know your way around it. It certainly doesn't help that the codebase is constantly changing as new features are added, bugs are fixed, tests are written, algorithms are optimized, and engineers come and go. Part of the system you understood months ago might have been refactored since then and now works completely differently. You'll always be chasing a moving target, so don't beat yourself up if you don't understand every corner of a codebase.

The best thing to do is to accept that you won't have a deep understanding of every single part of a codebase, and that's okay. As long as you work hard to form a mental model about the parts you're responsible for, things will start to make more sense. It won't happen all at once, but given enough time, the picture will become clearer and clearer. The trick is to be patient and get comfortable with reading unfamiliar code, because you'll be doing it for your entire career.

🔗 **RESOURCES**

- Learn to Read the Source, Luke[53] (blog.codinghorror.com)
- How to Read Code (Eight Things to Remember)[54] (spin.atomicobject.com)
- On navigating a large codebase[55] (blog.royalsloth.eu)
- How to read a code[56] (iamjonas.me)
- Reading Code is a Skill[57] (trishagee.com)

9 How to Add Value

As software engineers, we often get caught up in the day-to-day details of our job without even knowing it. We make hundreds of decisions each day, such as the architecture of our programs, what to name our variables, when to add a new function, which ticket to work on, how to design our database schema, and so much more.

While these are all fun decisions to make, they require us to consider the long-term implications of our choices, debate the pros and cons, and ultimately settle on a solution. There are so many choices to make that sometimes we fail to see how an individual decision fits into the grand scheme of things. We lose sight of the bigger picture because we're so focused on the details of the current problem we're trying to solve.

As you gain experience and progress in your career, you'll learn how your decisions fit into the overall system, and your decision-making skills

---

53. https://blog.codinghorror.com/learn-to-read-the-source-luke/
54. https://spin.atomicobject.com/2017/06/01/how-to-read-code/
55. https://blog.royalsloth.eu/posts/on-navigating-a-large-codebase/
56. https://www.iamjonas.me/2020/08/how-to-read-code.html
57. https://trishagee.com/2020/09/07/reading-code-is-a-skill/

will evolve. You'll start to comprehend the trade-offs between solutions and understand the positive and negative impacts your decisions could have on the business. You'll start to understand the implications of changing one part of the system and how it affects other parts. Eventually, you'll improve your ability to know which decisions add the most value to the customers and the business, and to prioritize those decisions above the others.

A strong quality of senior developers is that they're able to identify *on their own* where they can add value to the business. So, how do they do that? Let's take a look at different ways in which you can add value, both with code and without.

9.1 *Improve Productivity*

An important thing to remember as you work towards a senior developer role is that your job is not to build perfect software, which is a common misconception among developers when they're starting their careers. Yes, writing clean, reliable, and bug-free software is an ideal outcome, but ironically, that's not what you were hired to do. Your real job as a software engineer is quite simple—to build value for your company. Sometimes that means shipping code that isn't fully polished, and sometimes it means taking your time to get your design right before shipping a solution. In the end, it's all about shipping code. The more code you can ship, the more value you can deliver and the more valuable you become to your employer.

It's essential to understand the business and the industry you work in because you can't deliver disruptive value unless you have a deep understanding of your customer's pain points and how your products solve them. It may take a long time, sometimes even years, to fully understand the complexities of a business or an industry, but that domain knowledge is needed in order to deliver value. When you're just starting a new role, you'll only have limited experience, but there are still ways you can build value for your business that don't require years of institutional knowledge, and it's all about improving productivity so you can ship code faster.

9.1.1 IMPROVE YOUR OWN PRODUCTIVITY

There will always be one thing that you have full control over, and that is your own productivity. You may not fully understand the business context behind the changes you're asked to make, but at least you are able to control how quickly you can ship code and deliver those changes to customers.

Even if you're just getting familiar with a new codebase or learning a new language, there are habits you can build that will help you stay focused and work efficiently. Let's look at different ways to help increase your own productivity.

9.1.1 Minimize Distractions

It sounds obvious, but sometimes it's easier said than done. During the workday, do your best to stay focused on the task at hand, rather than checking news websites, your stock portfolio, or browsing social media. There are tools you can use that allow you to block certain websites or apps for a specified period of time so you're not distracted. And there are time management tools like the Pomodoro and Flowtime techniques that are designed to help you stay focused for short bursts of time.

9.1.1 One Task at a Time

It's easy to feel like you always need to be working on something, whether it's coding a new feature, fixing a bug, or refactoring some tech debt. We feel productive when we're sitting in front of our IDE churning out code—it's where we're most comfortable. It may be tempting to pick up a new task as soon as you submit your code for review, but sometimes that can hurt your productivity. You may think you can get started on another feature while you wait for your coworkers to review your code, but keep in mind that you'll need to switch context if you need to answer any questions or make changes to your pull request. This often can slow you down as you shift gears to focus between multiple things.

In the first few years of your career, you should try to focus on one task at a time, even if that means you're sitting idle waiting for feedback. There are other ways you can be productive while you wait, such as reviewing other developer's code, writing documentation, and planning and researching tickets. Remember, it's more efficient to give 100% of your effort towards one task than it is to give 50% of your effort towards two.

9.1.1 Use a To-Do List

There's a lot of information thrown at you every day. You may be focused on the task at hand, but your boss might ask you to do them a small favor. Or you might receive an email or chat message that you can't respond to right away. Or maybe your coworker assigned you as a reviewer on their PR that they're waiting to merge in. It's easy to let things slip through the cracks and forget about all the small things you need to do during your day.

One of the best ways to make sure you're keeping track of everything you need to do is to keep a to-do list. It sounds simple, but it's effective. A lot of developers try to keep track of everything in their head, and they forget to do certain things they promised or were asked to do. When you can write down your tasks on a notepad or an app, it helps you clear your head to focus on the task at hand. When you have a break in the day, make sure to check your to-do list for the next thing you should work on.

9.1.1 Plan Before You Implement

It's tempting to jump right into a task as soon as you pick one up, but it pays to be patient. Developers early in their career often have a habit of starting to code right away without thinking through the problem first. They may start implementing a solution without thinking about trade-offs or alternative solutions. It's a common mistake engineers make, and it can lead to reduced productivity if they need to change direction after the code has been written.

The saying "measure twice, cut once" is a good mantra to keep in mind when picking up a new ticket. Before you start coding a solution, make sure you've thought through multiple solutions so you can be sure you're implementing the correct one. This helps reduce the amount of wasted work if your first solution won't work out for any reason.

These are just a few examples of things you can do to increase your personal productivity. Every developer works differently, so try out different tools and techniques to find what works best for you. Once you've found your own productivity groove, the next step is to focus on helping increase your team's overall productivity.

- Attention Is My Most Valuable Asset for Productivity as a Software Developer[58] (zwbetz.com)
- Pomodoro Technique[59] (wikipedia.org)
- Flowtime Technique[60] (lifehack.org)

9.1.2 IMPROVE YOUR TEAM'S PRODUCTIVITY

Team productivity is one of the most important things an engineering manager focuses on. It's their job to make sure their team is working at as close to full capacity as possible. That means making sure their team members are not blocked by any issues preventing them from finishing their task, making sure the product requirements are fleshed out enough for the engineers to turn them into technical requirements, and making sure there is enough work coming down the pipeline for their engineers to pick up. The last thing an engineering manager wants is for their team members to be sitting idle and not shipping any code.

As an individual software engineer, there are things you can do to help your manager increase team productivity. As mentioned in the earlier section, Working with Your Manager,[§5] you should consider your manager's goals to be your own goals, so if they're concerned with keeping team productivity high, so should you be. If you can demonstrate to your manager that you can help them keep the team moving forward, you will prove that you're a valuable asset to the team and the organization.

So, what can you do to help improve your team's productivity?

9.1.2 **Limit Work in Progress**

If your team doesn't already limit the amount of work in progress (WIP), you may want to suggest implementing that practice for your team. It may seem counterintuitive that limiting the amount of work being done will help your team move faster, but the idea is to eliminate bottlenecks in the pipeline. Your team's output is limited by the smallest bottleneck. This could be, for example, the number of engineers available to work, the

---

58. https://zwbetz.com/
    attention-is-my-most-valuable-asset-for-productivity-as-a-software-developer/
59. https://en.wikipedia.org/wiki/Pomodoro_Technique
60. https://www.lifehack.org/911487/flowtime

number of engineers that have time to review pull requests, or the number of test engineers that are able to test features for quality assurance. When work begins to back up at one of these bottlenecks, productivity grinds to a halt. Adding a WIP limit allows you to control when work is released onto the queue to be picked up by engineers, which helps ensure that work is always moving through the pipeline and not backing up at a bottleneck. Another advantage of WIP limits is that they encourage engineers to limit the number of tasks they're working on, which allows them to focus on finishing a task before picking up the next one.

> **⧉ RESOURCES**
>
> - Putting the "flow" back in workflow with WIP limits[61] (atlassian.com)
> - DevOps measurement: Work in process limits[62] (cloud.google.com)

9.1.2 Keep Tasks Moving

Most engineers early in their career focus solely on completing their own tasks, but a common sign that a software engineer is ready to move to a senior role is when they shift their focus to helping teammates complete their tasks. When you complete a task, you may have some free time to help others before moving on to your next task, so try to leverage this down time to help keep things moving for the rest of your team.

> **》 EXAMPLE**
>
> Here are some examples of how you can keep tasks moving:
>
> - Don't let pull requests build up. If there are PRs waiting for review, focus on those. The work has been done and the code is waiting to be merged into the release pipeline. Help your teammates by reviewing PRs so they can move those tickets off the board and pick up the next ones.
> - Ask your teammates if they need help. Someone might be stuck on a tough problem and could use some advice, or perhaps another engineer is having trouble getting their development

61. https://www.atlassian.com/agile/kanban/wip-limits
62. https://cloud.google.com/architecture/devops/devops-measurement-wip-limits

environment up and running. If another developer is stuck, try helping to get them unstuck before starting a new task.

- Help define the requirements for upcoming tickets. Take a look at the tickets in your team's backlog and check to make sure there is enough information in each ticket for the developer to complete the task. Sometimes, tickets make it into the backlog with only a sentence or two of what needs to be done, when there are often additional product or technical requirements that need to be sorted out before a ticket can be started. Does the ticket need mockups from the design team? Is the expected functionality spelled out in the acceptance criteria? Are there steps to reproduce a bug? Making sure each ticket has enough information will help your team so they don't have to track down the information after they've already started the task.

Entire books have been written on productivity, at both the individual and team levels. These examples just scratch the surface of what you can do to help improve productivity, but if you want to add value for your team, there are always things you can do to increase the throughput for the work you need to deliver.

9.2 *Solve Pain Points for Your Customers*

Think about your typical customer for a minute. When a customer buys your product or service, they're looking to solve a specific need, whether that's to save time or money, perform a difficult task, automate something, or even provide entertainment value. What they care about the most is that their needs are met by your product—that's what makes your product valuable to them.

What they almost never care about is how your product is built. They don't care what language you used or what data structures, algorithms, or application architecture you chose for your codebase. They may not even care about all the features that are available in your product. What they do care about is that your product solves their pain points, which is why it's important to focus on the customer when writing software. Sometimes using a new framework or using fancy data structures is fun, but that doesn't always add value for your customers. Solving a customer's problems *always* adds value.

> ❯❯ EXAMPLE

Here are some good questions that will help you focus on your customers:

- How am I adding value for the customer right now?
- Which customers, if any, are requesting this feature?
- Which customers are affected by this bug that I'm fixing?
- Is this feature a must-have requirement before customer ABC
  signs the contract?

If you ask yourself these questions regularly as you work on new features and projects, you'll build a better understanding of how your work
adds value for the customer. You'll get a better idea of which tasks are
important and which are lower priority. Sure, there may be some tech debt
in a part of the codebase that's been bugging you for a while, but if fixing
it doesn't add any value for the customer then maybe you should focus on
something else that does. If you're adding value for the customer, you're
adding value for your company, and in the end, that's all that matters.

9.3 Ship Code That Increases Revenue

It's satisfying to see your code in the hands of customers and providing
them value, especially when they pay for your product. It's even more
fulfilling to write code that has a high impact, especially when it comes
to helping your business grow. If you're lucky, the code you write may
become necessary to the success of the business, at which point it
becomes *business critical.*

Business-critical code is typically complex because it's required to
handle multiple use cases for many different customers. The complexity
grows over time as new use cases are discovered and new customers are
supported. However, business-critical code doesn't necessarily equate to
state-of-the-art code. Surprisingly, the most profitable code is sometimes
rather outdated and boring. The most important part is that the code
works, is reliable, gets the job done, and is in the hands of the customer.

All business-critical code becomes legacy code over time. It's just a
fact. When you have a codebase that's necessary to the success of the business, making changes to the code can be risky if not planned carefully.
Small improvements add up over time, but you probably won't be refac-

toring half the codebase to a new architecture. As they say, "If it ain't broke, don't fix it."

As long as the critical systems are in place and still making money for the business, there will always be a need for people to maintain and support those systems. Depending on the expertise and domain knowledge required, you may be able to leverage your knowledge of the codebase for job security and a rewarding salary.

Keep in mind there are trade-offs, as with most things related to software engineering. Although job security is important to a lot of people, you should expect to be working on possibly outdated technology and working with code that other people wrote. Legacy code can often be difficult and time-consuming to change and stressful to debug. It's certainly not for everyone, but some people prefer to work on legacy systems rather than the ambiguity of newer codebases.

With most legacy systems, things move slowly. You'll need multiple approvals to make any significant changes to the system due to the risk involved. While some developers don't mind the slower pace, others may find it frustrating to get anything done. You may be better suited to work on projects where you're building new products or iterating on existing ones, especially early in your career.

A common misconception among new developers is that you'll be working on interesting problems and engineering state-of-the-art solutions all the time. While this may be the case for some products, more often than not, the code that makes money will be rather boring.

> ⟫ EXAMPLE

> A checkout page for an e-commerce company is one example of code that is almost never state of the art. It's a well-known problem with well-known solutions, and it's not exactly the most exciting part of the codebase. But every single transaction in an e-commerce business flows through that page, so the code that powers that checkout page is code that makes money.

9.4   *Ship Code That Reduces Cost*

While somewhat counterintuitive, a common path to adding value is to ship code that reduces cost for your customers or for your company. If a product solves a customer's pain point better, faster, or cheaper than if they built it themselves, the customer benefits from a return on their investment. When a customer purchases your company's product, they're doing it in order to free up capital or other resources. This allows them to shift resources to help grow other parts of their business. It's a win-win for both your customers and your company.

While saving your customers money is a great way to add value for your company, writing code that reduces cost for your *employer* is another opportunity for you to have a significant impact. Operational efficiencies are vital in highly competitive industries, as that can make the difference in whether the company makes or loses money in a given quarter or year.

Businesses are incentivized to control their costs and increase their efficiency as they work to produce more goods and services with less resources. This usually means building internal automation tools, scaling existing infrastructure, and streamlining processes to gain efficiency. While you may not get to work on projects like these right away, there are things you can focus on that help reduce the cost of delivering software, such as writing clean and modular code.

Clean, modular, and extendable code adds value by reducing the amount of time it takes to make future changes to the codebase. Business requirements always change over time, and that requires frequent modifications to the codebase to support the requirements. Let's look at some ways in which you can add modularity and extensibility to your codebase so you can ship more code with less cost.

9.4.1 CONFIGURABLE CODE

Different customers have different requirements, so ideally, we'd write software that can be configured to each user's individual needs and preferences—all while using the same codebase. It's generally considered a bad practice to hard-code configuration values because, at some point, you may need to change that value. If it's hard-coded into the codebase, that means you'll need to modify the code and deploy a new release in order to change the configuration, which is time-consuming and adds risk.

When we hard-code logic for specific configurations in our codebase, programs become rigid and inflexible, making it difficult to modify the system as our customer's requirements change. Therefore, it's better to build our systems in a way that can easily handle changing requirements.

Whether you need to add a new shipping option, toggle a feature on or off, or increase the threshold for some limit, try to build it in a way that lets you or someone else update those values without needing to deploy new code. When you separate configuration values from the application itself, you're able to modify the software to fit customers' needs with less time and effort.

Existing applications probably already have some sort of configuration system, so it's important to understand how it works. Try to learn what its possibilities are and also where the functionality falls short. In almost all cases, it's better to use the existing configuration system rather than build your own from scratch.

9.4.2 **ABSTRACTIONS**

It's often difficult for new programmers to think in abstract concepts when writing code. It often comes down to the fact that you don't know what you don't know, and that's okay. Sometimes it's hard to think about how to abstract code if you've never seen it done before. Luckily, an easy solution to build this skill is to study other people's code.

The single best way to learn abstractions is to study code written by experienced engineers. An experienced engineer will write code abstractions in order to be able to adapt that code to many different use cases. By writing code that is abstract, they can easily extend or reuse the logic as requirements change because the code is designed to be used in multiple scenarios.

A well-abstracted part of a codebase allows teams to make changes quickly when they need to support new requirements for the business. This allows them to deliver value more efficiently, which saves the company money and allows the programmers to move on to the next task.

Unfortunately, there's no silver bullet to learning this, as it is both an art and a science. Over time, you'll learn how to identify opportunities where code can and should be abstracted to handle different use cases.

The more experience you gain during your career, the more natural it'll be to design your programs with abstractions in mind. You'll see how other developers separate their logic, which will influence how you write your own code. Writing modular code is a skill that takes years to develop, but finding time to study code written by experienced engineers or in popular open-source projects will help you learn and identify patterns you can reuse. You'll pick up techniques and start to incorporate them into the code you write, and soon enough, you'll be able to write abstracted and modular code.

Sometimes, small abstractions can have a big impact on how quickly you're able to extend the logic, saving you time and resources, so it's worth taking the time to build that skill. But be careful, though. As you learn to think about abstractions and build them into your codebase, keep in mind that every new level of abstraction added will increase complexity and the cognitive load you'll need to fully understand how your program works. Additional levels of abstraction also often come with a performance cost, so it's important to consider the trade-offs involved with introducing new abstractions in your code.

9.4.3 WRITE AUTOMATED TESTS

For most engineers, writing automated tests feels like a chore at first, but over time, you'll understand why automated testing is critical in the software development lifecycle. There are many benefits to building out good code coverage with a test suite, but the most important reasons we'll look into are:

- To catch bugs earlier in the process before they hit production.
- To reduce the burden on QA engineers from having to manually test each part of the code to ensure the next changes don't introduce regressions.

That's it right there. Those two bullet points address issues that are incredibly costly to an engineering organization, so if you can ensure all of the changes you make have corresponding test coverage, you'll be able to ship code that helps reduce costs for your company. You'll know your code works according to your test cases, so the QA engineers can focus on testing other parts of the application or building test automation.

A bug is cheap and easy to fix in your local environment, but it becomes exponentially more costly to fix a defect that's moved through staging environments and been deployed to production. The cost is not only in the time and resources needed to identify, fix, test the fix, and deploy the updated code to the affected environments, but also in the opportunity cost of not being able to work on another task that could build value, not to mention any cost of mitigating damage done by the bug. If you can reduce the amount of bugs introduced into staging and production environments, you can reduce the cost associated with shipping code.

Building value with an automated test suite is similar to forming a new habit—it's going to require some real effort, and it's not going to happen overnight. But if you chip away at it with every code commit, you'll eventually have good code coverage on all new code written, and you'll hopefully build a healthy culture around testing within your team.

Having an automated test suite with good code coverage improves the efficiency with which you can confidently make changes to the codebase. You'll be able to quickly refactor and make modifications to your business logic while knowing that you're not breaking existing functionality. Software is complex, and there are often unintended consequences from our code changes. A test suite with sufficient code coverage will help you catch those side effects before they make it to staging or production, leading to higher-quality software and more time to work on tasks that build value.

An important point to understand, however, is that your tests are only as good as the assertions they make. Your tests can only catch behavior that you're explicitly testing for, so don't expect your test suite to catch every bug for you. Tests can be incomplete, and they can have bugs, just like all code.

◇ CAUTION   Just because your tests are passing does not mean your code is error-free.

Now that we've covered how tests help reduce the cost of shipping software, let's look at *how* and *when* to write tests.

9.4.3 Fixing Bugs

Any time you're fixing a bug, include at least one unit or functional test with your code changes if you can. The benefits are twofold—first, you're adding a test to the suite that ensures that the bug is in fact fixed, because you can set up a test case with steps to reproduce the bug and to prove with assertions that the bug no longer exists. Second, you now have tests written that will catch that bug in the future in case you or another engineer make changes that accidentally reintroduces it.

9.4.3 Refactoring

The complexity of a codebase increases over time. As new developers join the team, experienced developers move on to new roles. Business requirements change and parts of the codebase will need to be rewritten. The more lines of code you're rewriting, the riskier the refactor becomes. So, how do you replace old code in a production environment without breaking anything?

Automated tests are not the silver bullet, but they're a crucial tool that can be used to ensure you're not breaking existing functionality. When refactoring code, the goal is to make sure you have tests in place *before* you begin making changes to the structure and logic of the code.

This helps define the existing functionality: how the system should work. As you refactor your code, run your test suite periodically to ensure that any new changes you make aren't breaking the expected behavior. If your tests are still passing, you'll be confident the system is still working as intended with the modified code.

9.4.3 New Feature Development

It's always fun to write something new rather than fix or extend code that someone else has written. You get to choose the names for your methods and variables, design the APIs, and come up with a clean solution. As you're building out new features in your codebase, think about how you're going to test the code that you're writing.

Ideally, all new features you add should include automated tests. This isn't always possible, but try to limit the situations in which you commit code that doesn't have associated test coverage. If for some reason the code is difficult to test, try asking yourself how you can modify it to make it easier to test in the future.

9.4.4 DOWNSIDES OF TESTING

A proper automated testing pipeline and a test suite with good test coverage adds a significant amount of value and can reduce the amount of cost associated with writing software. But automated testing does not ensure you're producing quality software. There are trade-offs you should be aware of when writing tests.

It's entirely possible to write bad tests. Some of the features we build and bugs we fix are complicated. You may find yourself writing dozens of lines of code to set up the data for a test, which introduces opportunities for bugs within your tests.

Here are some examples of how tests can be bad:

- **Incorrect tests.** These are tests that test the wrong behavior of a piece of code, or tests that pass when they should actually fail. This can be misleading since you think your code is well tested when in fact it isn't.
- **Poor code coverage.** The tests do not cover all of the code paths or edge cases needed in order to be confident a piece of code works the way it should. If there are parts of a system (whether a few lines in a function or entire files in a module) that are not covered by tests, you may be missing critical bugs.
- **Poor maintainability.** If you have tests that frequently fail or fail inconsistently when other parts of the system change, your tests will be difficult to maintain. Additionally, tests need to be updated as the underlying code they're testing is modified and refactored. Tests are code, after all, and will need to be maintained along with the core business logic.
- **Nondeterministic tests.** These are tests whose behavior is not consistently reproducible. For example, if a test generates random test data, then your tests will not run under the same conditions every time. Some data may cause a test to fail, but it will be hard to reproduce because the test will generate different data the next time it runs.

Designing good tests takes some skill, but with a little practice, you can learn good testing habits in no time. The main idea is to build towards correctness and determinism, then increase coverage and maintainability as much as possible.

Here are things you can do to ensure tests are isolated and consistent:

- Make sure your test data is the same every time. It's okay to generate fake data dynamically, but be aware that test data that changes for each run may introduce unreliable results, causing your tests to fail intermittently.
- Make sure your tests do not rely on shared state, such as data from a database or cache that other tests are also modifying. Ideally, each test sets up its own data and cleans up the data after itself.

- Each test should have one responsibility and should be testing one thing. Tests that make multiple assertions are more prone to being unreliable.

It's important to remember that faulty or incomplete tests can actually have a negative impact on productivity for you and your team. All code you write needs to be maintained, even the tests. When a test breaks, it'll need to be fixed, which means taking time out of your day to track down the failing test, reproduce the failure, and fix it. And any time you spend fixing a broken test is not time you're spending adding value for your company and your customers.

It takes a fair amount of work to build and maintain a test suite with good code coverage, but don't let the amount of work deter you from doing it. Anyone with a little discipline can build a valuable test suite, but sometimes you'll need to hold your teammates accountable when they try to commit code without test coverage. As with all good habits, it will take time at first to build a foundation, but once it's there it'll help reduce costs over time.

You may be fortunate enough to work for a company or a team that has a dedicated team of test engineers that maintain some or all of the automated test suites. While having a team responsible for testing code quality is helpful, it does *not* mean you are free to add features and make changes without writing tests for your own code. All engineers on a team share the burden of making sure the code works correctly, so it's important to work together with other developers and test engineers to ensure good code quality. You will most likely still be responsible for writing unit or functional tests, and the test engineers will be focused on writing integration and end-to-end tests for the system as a whole.

🔗 RESOURCES

- The exponential cost of fixing bugs[63] (deepsource.io)

9.5 *Add, Improve, or Automate Processes*

While some people may embrace a well-defined process that provides structure, it may bring a negative connotation to others who value auton-

---

63. https://deepsource.io/blog/exponential-cost-of-fixing-bugs/

omy and independence. At some point, all programmers feel the frustration of dealing with "red tape," or the lengthy process to gather the necessary approvals to move forward with a decision. But processes aren't always a bad thing, as they provide a tremendous amount of value.

Whether you work for a large Fortune 500 enterprise, a development agency, a small scrappy startup, or even freelance for a living, you will follow processes every day and probably even develop new processes. Building a process provides value because it allows you to standardize a set of steps to complete specific tasks. By doing so, you are able to increase efficiency and scale your throughput by training others to follow the process.

When you take a set of steps, which may need to be done in a specific order, and formalize them into a process that can be followed by anyone anytime that task needs to be completed, you add value by creating *consistency*. You've defined a standard way of doing something that can be completed the same way every time, regardless of who is performing the task.

Assuming you've documented your newly formed process, you unlock numerous advantages. First, you can reference the steps in the process at any time, which can save you in the future if you forget how to perform a task you haven't done in a while. Sometimes a process only needs to be performed on a monthly, quarterly, or yearly basis. As long as you have your process documented, you can always reference the documentation and be confident you're performing the steps correctly every time.

Second, you can scale your process by training other people on how to perform the tasks. Whether it's a sales process or a deployment process, you're now able to train an entire team on how to perform a specific task. You have just created value.

Furthermore, processes can be measured, and anything that can be measured can be improved. If you're able to define specific and quantifiable metrics that accurately reflect the efficiency of the process, you can track how well you're performing the process. Once you have a baseline for how efficient your process is, you can work towards refining it to improve that efficiency even further.

And lastly, in many cases processes can be partially or fully automated. Processes often define a set of repeatable steps, and if some or all of those steps can be automated with scripts, programs, or third-party tools, you may be able to reduce the burden on teams and individuals who rely on those processes. The time saved by automating a process adds up over time and can provide a significant amount of value for a company depending on how often that process is performed.

So, as a developer, how can you add processes to add value? Start by observing your day-to-day tasks, and look for things you repeat over and over again. A good rule of thumb is if you repeat a task more than three times, that's a potential candidate for a process, if there's not already one in place.

>> EXAMPLE

- Help your team formalize an onboarding process for new hires.
- Help formalize a deployment process to ensure the code getting pushed to production has been thoroughly tested.
- Add a process around code reviews, even if it's just adding a standard template for the description to include testing instructions, a change log, multiple approvals, or anything else that should be filled out before opening up a pull request.
- Propose adding, modifying, or removing steps in the software development lifecycle that you think will help your team work more productively.
- Push for better-defined requirements from the product team before you begin designing the technical solutions.
- Push for more-thorough design documents before you and your teammates begin implementing a solution.
- Document the process for applying a hotfix to a release build.
- Document the steps needed to manually run a set of commands from the command line to perform a specific task.
- Help formalize a bug reporting process to make sure you're gathering the required data every time, such as date first seen, browser version, the user's role, and steps to reproduce.

With a little creativity, you can create a process around almost anything. And as you build out these processes, you'll start to see what works and where the processes break down and need to be improved. Then, you can rework them to accommodate a new use case and improve the process. This improves efficiency and reduces errors because it's documenting a standardized way to perform a task. Think of it like bumper lanes in a bowling alley or swimming lanes in a lap pool. The process is supposed to guide someone to perform a task while reducing the chances of making costly mistakes.

Processes are a great way to formalize a set of steps that others can follow. Processes give you leverage to scale those tasks at the expense of adding rigidity and extra work. Keep in mind that too much process can be a bad thing, so there's a balancing act you'll need to look out for if you're looking for areas where you can add or improve a process.

9.6 *Add Value without Writing Code*

While most of the ideas in this section so far have involved coding, there are other ways to add value without writing any code, such as writing documentation. Often overlooked by both young and experienced developers, the ability to write clear and concise documentation will set you apart from the rest of your teammates.

Transferring knowledge through written sources has been used by almost every civilization that's ever existed. There's a reason it's so valuable, and writing is a skill that anyone can do and anyone can learn.

As a business matures, knowledge is gained. That knowledge is valuable, and it probably took a lot of effort on someone's part to build an understanding of a specific problem, a customer pain point, or a nuance of the industry. When you work hard to solve a problem, fix a bug, or develop a new feature for a customer, try to document what you learned on an internal knowledge base.

It may seem like it's extra work, but it adds significant value even if it goes unnoticed at the time. For starters, you can reference this in the future, as mentioned earlier. You may forget why you made a certain decision or chose to solve a problem in a specific way, and it's helpful to be able to look back on your notes to figure out the context behind a decision.

Second, it's an easy way to contribute to the knowledge base for your company, and it's something you can do on *day one* when starting a new job. You may not even have access to the codebase yet, but you can make edits to the onboarding instructions if something was confusing or missing.

Third, the content you create on your company's knowledge base will be searchable, so it will help your teammates find answers to their questions quicker and sometimes without interrupting you and your coworkers. It's an asynchronous way to share knowledge, ideas, decisions, and processes that help newcomers build on the work of those before them.

Fourth, new developers will join your team and experienced ones will leave. Sometimes, important knowledge about how part of the codebase works or why something was built the way it was will disappear when a developer leaves the team. By documenting what you know and what you learn, as well as encouraging others to do the same, you're able to pass along that knowledge to those who come after you.

Just like well-defined processes, thorough documentation is a quality of high-performing engineering organizations. If your team already has good documentation, you can add value by contributing to what's already there. If your team doesn't have any documentation, you can add value by documenting what you know and what you learn as you fix bugs and design new features, and by sharing that information with your teammates.

An important point to remember, however, is that it's common for documentation to become out-of-date if it's not updated regularly. When you're concerned with shipping a new feature or a critical bugfix, it's easy to forget to update the documentation to reflect the new changes. It takes discipline to keep documentation up-to-date and prevent it from becoming a knowledge graveyard. If you don't take time to remove outdated documentation, it could easily have the effect of turning people away because they know the information in the company wiki is out-of-date and no longer relevant. Just like code, it takes a lot of work to maintain an internal knowledge base, but the payoff from a well-maintained system is significant.

9.7 *Wrapping Up*

As you can see, there are so many ways you can add value to make yourself, your team, and your company more efficient. It'll take a little creativity at times, but the opportunities are there if you can find them.

The techniques outlined in this section only scratch the surface of ways in which you can add value, so you should constantly be looking for new opportunities. In doing so, you'll prove to your manager that you're a valuable asset to the team and that you're constantly putting the team first while trying to improve the system. A rising tide lifts all boats, and you have the power to drive improvements within your organization. Now it's up to you to find those opportunities.

🔗 **RESOURCES**

- Code-first vs. Product-first[64] (thezbook.com)
- What is value?[65] (dwyer.co.za)
- You Are Not Paid to Write Code[66] (dzone.com)
- How Can You Increase Your Value as a Software Engineer[67] (tomastrajan.medium.com)
- How to increase your value as a developer[68] (enterprisecrafts-manship.com)
- The Software Engineer's Path to Providing More Value and Gaining More Visibility in Your Organization[69] (bytepercep-tions.com)

10 How to Manage Risk

In the previous section,[§9] you learned that one of your primary responsibilities as a professional programmer is to create value. We looked at a number of ways you can create value, both with and without writing code. In this section, you'll learn about a second major component of your job, which is to manage risk.

Before we jump into the details of how to manage risk, we need to take a step back and answer a fundamental question: what is risk?

Risk in software development is the probability for uncertain events to occur that can have a negative impact on an organization or other entity. In essence, risk is the probability for bad things to happen. Shipping buggy code, lack of communication, changes to consumer behavior, and missing deadlines due to poor project planning are all kinds of risk that can harm

---

64. https://thezbook.com/code-first-vs-product-first
65. https://dwyer.co.za/what-is-value.html
66. https://dzone.com/articles/you-are-not-paid-to-write-code
67. https://tomastrajan.medium.com/
   how-can-you-increase-your-value-as-a-software-engineer-cab9599bbbe
68. https://enterprisecraftsmanship.com/posts/how-to-increase-your-value-as-developer/
69. https://www.byteperceptions.com/marketing-for-software-engineers/
   strategies-to-get-a-promotion.html

your organization. Risk comes in all shapes and sizes, and each one has the potential to cause harm in one form or another.

As software engineers, we're focused on building scalable software systems, but what many developers lose sight of is that we're also responsible for keeping those systems up and running and ensuring a seamless experience for our customers. Software has become a critical component of everyday life, and so managing risk has become a critical component of software engineering.

Risk is always there. That's a fact. So, it's better to understand it and embrace it rather than ignore it. The more senior you become in your career, the better you will get at identifying and planning for risk. Managing risk is one of the most important skills you will use in your software career.

It's important to note that some risk is acceptable. Senior software engineers learn how to manage it, not eliminate it. It takes time for junior engineers to learn how to identify different areas of risk. And as with most things in software engineering, there are trade-offs. Senior engineers learn when to allow low-probability or low-impact risks into the system because it allows them to move quickly.

Before we get into details about major risks involved in software engineering, let's look at the different kinds of risks you may encounter.

10.1 *Types of Risk*

Unfortunately, because a lot of risk management comes down to experience, you won't learn everything there is to know about it in this section. What you will learn, however, are the different types of risks so that you are aware of what to look out for in your day-to-day roles.

- **Technical Risks**
  - Poor code quality
  - Poor technical decisions that prevent you from adapting to changing requirements in the future
  - Lack of documentation and knowledge sharing
  - Poor technology choices
  - Poor code performance

- **Scheduling Risks**

  - Poor project estimation
  - Requirements that are not finalized and keep changing
  - Lack of visibility into the work that is in progress and what has been completed
  - Project scope creep

- **Operational Risks**

  - Poor communication
  - Lack of proper training
  - Lack of effective processes or procedures
  - No separation of concerns, or checks and balances
  - Data breach due to poor security practices

- **External Risks**

  - Sudden market changes
  - Increasing competition
  - Government regulations
  - Changes to consumer behaviors
  - Weather and natural disasters (yes, really)

This is by no means a comprehensive list of every risk involved in software engineering. There are many additional categories and subcategories of things that can go wrong while keeping software systems running. Luckily, you're not the only one responsible for managing these risks. Risk management is a team effort, but that doesn't mean you're off the hook when it comes to doing your part.

> 𝒫 RESOURCES
>
> - Summary: Waltzing with Bears — Managing Risk on Software Projects[70] (geoffmazeroff.com)

---

70. https://geoffmazeroff.com/2017/12/03/
    summary-waltzing-with-bears-managing-risk-on-software-projects/

10.2 *Contributors to Technical Risk*

Let's look into some of the major contributors to technical risk that you should look out for in your day-to-day role.

10.2.1 OVERENGINEERING VS. UNDERENGINEERING

Effective engineering is about shipping software quickly while preserving your ability to make additional changes quickly in the future. The goal is to move fast without putting yourself in a situation you'll later regret. In essence, we need to build software that meets the current requirements for our customers but leaves enough flexibility to easily extend the code to handle additional requirements in the future.

Seems easy, right?

The longer you work as a professional programmer, the more you will come to realize that good code approximates the complexity of the problem at hand. Good code is not needlessly complex, but not overly simple either. The best engineers are able to design and build solutions that match the complexity of the problems they're solving.

However, a lot of software engineers early in their career don't have enough experience to know how to match their solutions to the complexity of the problem, so they end up either underengineering or overengineering their solutions. There's no simple answer as to how to avoid these situations, unfortunately, but just being aware of each one is a step in the right direction. Over time you'll naturally gain an understanding of when a solution is being under- or overengineered. In the meantime, let's look at each one a little deeper so you can better identify each situation.

10.2.1 **Underengineering**

When a developer underengineers a solution, they are not doing enough forward thinking when designing a solution to a problem. Although they may be focused on solving the immediate problem at hand, they may be losing sight of a better long-term solution. This tends to be a common trait among developers who are just learning how to write code, because most of their energy is spent on getting the program to work. Once they come to a working solution they move on to the next task. That can cause problems in the future.

Just because a piece of code works and compiles without errors doesn't mean it's ready to ship. There may be better ways to solve a problem that allow for more functionality in the future. While the original solution solves the problem right now, the code may need to be significantly refactored when it needs to handle additional use cases in the future.

Underengineered solutions often contradict the Don't Repeat Yourself (DRY) principle. The DRY principle is a common guideline software engineers use while structuring their code so they are not repeating the same logic in different parts of the codebase. This is good because it encourages programmers to structure their programs so the logic can be written once and reused in multiple places throughout the codebase.

When you follow the DRY principle, you can often add additional functionality to your code with little effort because you only need to make changes to one part of the codebase when updating logic. Additionally, when updating logic that is repeated throughout the codebase, you risk the chance of missing a block of code. This increases the possibility of introducing bugs into the system during refactoring and may lower the quality of the codebase over time.

A common rule of thumb is if you're noticing yourself copying and pasting blocks of code throughout your codebase, that could be a sign that you need to consolidate your logic so that you're not repeating it. It's a simple technique that goes a long way to reducing the amount of risk involved in making future changes to logic.

Underengineered solutions also sometimes contradict the Single Responsibility principle, which states that modules, classes, and functions should have only one responsibility over a program's functionality. If you find yourself writing a class or a method that's doing multiple things, such as calculating values, transforming data, and storing it in a data store, then you may want to rethink how your solution should be designed.

Underengineered solutions tend to try to do everything in a single class or function, when they really should be broken up into multiple pieces that each handle a separate task. Solutions that contradict the Single Responsibility principle tend to be difficult to extend and often need to be refactored when new functionality needs to be added. Just like the DRY principle, following the Single Responsibility principle is a simple technique that will reduce the risk of needing to rework the code in the future.

10.2.1 **Overengineering**

On the other end of the spectrum, overengineering is the act of designing an overly complex solution to a problem when a simpler solution would do the job with the same efficiency. Software engineers often fall into this trap because they add unnecessary complexity to the system *just in case* it will be needed in the future. In essence, it's the act of solving one problem while optimizing for other requirements that don't, and may never, exist. When developers overengineer solutions, they're often thinking about theoretical scenarios that could come up in the future but are never guaranteed to happen, which leads to extra time and energy spent writing, testing, and debugging code that isn't required.

When you end up with code and logic in your system that is overengineered, it increases the difficulty of reading, understanding, and modifying the code for your teammates. Developers will need to work around the complexity in order to add enhancements or fix bugs.

Plus, overengineering a solution directly contradicts the Keep It Simple, Stupid (KISS) principle, which argues that most systems will work best if they are kept simple rather than made complicated. If you strive to write code a junior engineer will be able to understand and modify, you're probably in good shape. If you add unnecessary abstractions or try to be clever with your solutions, you're probably not thinking about the risk of later developers modifying your code without fully understanding what it's doing.

The production lifetime of the code you write will likely be years, and you and other developers will eventually need to revisit that code and modify it to add new functionality. Code that is less complex will always be easier for future developers to understand and refactor than code that is more complex.

From a risk perspective, overengineering a solution may hinder your team's ability to move quickly in a different direction in the future. Complexity often adds rigidity to code, because it is harder to refactor or modify when the business priorities change. Your goal should be to write clean and concise code, but not so clean that it constrains your ability to move and adapt in the future.

If possible, try to strive for the Goldilocks Principle—just the right amount of engineering and nothing more. Unfortunately, that comes with experience, and it's easier said than done.

🔗 RESOURCES

- KISS Principle[71] (wikipedia.org)
- Under-engineering, over-engineering, right-engineering[72] (blog.startifact.com)
- Stop Overthinking Your Complex Solutions and Start Building Simple Ones[73] (betterprogramming.pub)
- Overengineering: Why We Do It and 10 Ways to Tackle It[74] (betterprogramming.pub)

10.2.2 LARGE REWRITES VS. INCREMENTAL REFACTORING

As software developers our job is never done. There is always more work to do on the codebase, whether that's adding new features, cleaning up technical debt, improving performance, or maintaining a legacy system. At some point in your career, you'll be faced with the decision to continue adding to an existing codebase or to rewrite the system from scratch in a new project.

Both paths involve significant risks that it's good to understand before making any major decisions. When deciding whether to refactor a legacy codebase versus rewriting it from scratch, you should take a number of factors into account such as the type of application you're dealing with, your team's capabilities, the available resources, future hiring plans, and your organization's general appetite for risk.

Fortunately (or unfortunately), the decision is most likely not yours to make. The most senior engineers on your team will probably be the ones to make the decision along with your manager, because they will be the ones with the most experience and will understand the implications better than you will.

That shouldn't stop you from contributing to discussions and lending your opinion, however, so let's look at some of the risks involved in both paths.

---

71. https://en.wikipedia.org/wiki/KISS_principle
72. https://blog.startifact.com/posts/older/
under-engineering-over-engineering-right-engineering.html
73. https://betterprogramming.pub/
stop-overthinking-your-complex-solutions-and-start-building-simple-ones-712400ea8385
74. https://betterprogramming.pub/
overengineering-why-we-do-it-and-10-ways-to-tackle-it-460663d35ff3

10.2.2 Refactoring

If you choose to refactor a legacy system, you will be making incremental changes to the codebase to clean it up over time in order to get it to a more manageable state. The goal is to improve the internal structure of the code without altering the external behavior of the system.

Pros

- Doesn't divert resources away from legacy systems.
- Improvements can be isolated to specific parts of the codebase in order to limit the risk of introducing breaking changes.
- Always an option; you can refactor as much or as little as you want as you have the resources.
- Any codebase or architecture can be refactored incrementally.

Cons

- Limits you to working within constraints of the legacy system.
- While it improves code, sometimes you cannot fix underlying architectural issues.
- Often difficult and complex to untangle the web of legacy code.
- May require writing new automated tests prior to being able to refactor the business logic.
- Refactoring maintains the status quo, so it's difficult to introduce new features or functionality.
- Requires discipline to manage the complexity. The application will be in a transitional state as individual parts of the codebase are refactored.

10.2.2 Rewriting

The big rewrite happens when you start from scratch with a new codebase. It may sound enticing and straightforward, but the amount of work is almost always underestimated. This is often done concurrently with changing to a new platform, such as moving from on-premises servers to the cloud or moving to a new chip architecture as hardware is upgraded.

Pros

- Enables foundational changes to a part of the system, often introducing new capabilities thanks to new technologies or design decisions.

- Eliminates the need to retrofit old code to meet new use cases because you can build for them without any technical debt.
- Engineers are able to set new coding standards with a clean codebase.

Cons

- Always takes longer than anticipated, eating up resources for other projects and increasing the possibility that management will abandon the project.
- Not guaranteed to solve all problems that plagued the legacy system. Sometimes those are due to systemic or cultural processes rather than the technology or codebase.
- Complex migration periods as you phase out the legacy system.
- Duplicates the amount of work during the transition period. One team builds the new system while another continues to maintain the legacy system.
- Requirements for the new system are a moving target as the legacy system still needs to be maintained and upgraded. New functionality may need to be implemented in both codebases.

Every codebase is unique, and every business has different competing priorities, so the decision to refactor or rewrite an application is not a one-size-fits-all problem. You and your team will need to weigh the pros and cons and determine the risks involved in either choice before making a decision.

10.2.3 BYPASSING PROCESSES

In the previous section,[9] we discussed the importance of adding or improving processes, and how they add value to an organization. Processes give you guardrails that enable consistency and allow teams and organizations to scale and pass down business knowledge.

But not all processes are created equal, and sometimes processes can feel like they're getting in your way. A lot of developers don't want to deal with the "red tape" that processes add to the software development lifecycle, and most would rather just write more code instead of getting slowed down by seemingly unnecessary processes. Eventually, a developer may cut corners and break protocol.

⟫ EXAMPLE

Here are a few examples where developers sometimes bypass processes:

- They may merge code to the main branch without a proper code review because they don't want to wait for feedback, leading to a bug that could have been easily caught.
- They may elect not to use proper naming conventions because they don't want to take the time to search the docs to find out the correct way to name an environment variable, leading to a broken deployment because the code expected the variable to use a certain naming convention.
- They may do some work without creating a proper ticket in the bug tracking system, leading to changes that are hard to audit and track down.
- They may commit an inefficient SQL query without running an EXPLAIN on it because they think it's a harmless query, leading to a slowdown in database performance.

Yes, some processes can be frustrating, and it may feel like they're just slowing you down unnecessarily, but processes exist for a reason. When you bypass processes, whether it's on purpose or by mistake, you're actually introducing more risk that something in the system may fail.

Next time you find yourself frustrated and wondering why you have to follow a process, ask yourself why you think the process is there in the first place? What could go wrong if it wasn't followed? Hopefully, that'll help you understand and appreciate a little extra red tape here and there if it means saving you from making a catastrophic mistake.

10.2.4 SOFTWARE DEPENDENCIES

Almost every codebase leverages third-party libraries and external dependencies to provide some part of its functionality. Why reinvent the wheel and build a library from scratch when you can use an open-source package that solves the problem better than you ever could? Add the fact that thousands of other developers use the library and consistently file bugs and contribute fixes to it so it improves over time, and it sounds like a no-brainer, right?

Most of the time, utilizing third-party libraries saves you time and money because you won't need to implement and maintain a solution yourself. But be careful, because there is a hidden cost to any third-party library you pull into your codebase. Every time you add a new dependency to your codebase, you're introducing new areas of risk, because your system now relies on someone else's code in order to function properly.

Sure, you might be able to view the source code and gain confidence that the software does what it claims it does, but that's not the only kind of dependency risk you should be worried about.

>> EXAMPLE

Here are some other examples of dependency risks:

- **Security risks.** The third-party code that you add to your system may add new attack vectors to your codebase that you may be unaware of. Hackers often exploit known vulnerabilities in specific widely used libraries.
- **Upgrade risks.** The third-party code may change over time as they add new features and apply bug fixes. They may introduce breaking changes that in turn cause your own code to break after upgrading to a new version, forcing you to drop everything to fix new bugs that were introduced into your system.
- **Dependency graph risks.** You may be able to read the source code of your third-party dependencies, but those libraries may rely on their own dependencies, and those dependencies rely on their own dependencies, and so on. This creates a brittle dependency graph that can easily break your codebase. In some cases, it may be hard to remove or upgrade dependencies that have known bugs, because the library in question is a dependency of another library you installed, so you're at the mercy of your dependencies to fix the issues for you.
- **Supply chain risks.** Supply chain attacks are becoming more common in the software industry. Supply chain attacks occur when someone uses a third-party software vendor to gain access to your system. When you install third-party libraries into your codebase, you are granting that code access to your system. If an attacker is able to compromise a third-party library that has been installed on your system, they'll be able to access your data and possibly your infrastructure. Sometimes hackers will target little-known but critical libraries that

are deep down in the dependency graph, making supply chain attacks difficult to prevent and mitigate.

Hopefully that gives you a good understanding of how introducing third-party libraries into your codebase also introduces added risk. Next time you're searching for a third-party library, ask yourself if it's really needed. If the code is open source and relatively small, it may be better to study how it works and build your own similar solution. This is not always feasible, however, since some third-party libraries can contain tens of thousands of lines of code.

🔗 **RESOURCES**

- Supply Chain Attack[75] (wikipedia.org)
- Supply Chain Attacks: Examples and Countermeasures[76] (fortinet.com)

10.3 *Mitigations for Technical Risk*

10.3.1 DIVIDE AND CONQUER

People often compare the ability to program a computer with superhuman powers. Sure, it may seem like that at times when you see programs do things that are seemingly impossible or futuristic, but programmers are only human. There's a limit to how much the human brain can comprehend at any given time, and we often find that limit when learning a new codebase or managing a large project at work.

There are some projects that are so incredibly complex that they cannot be built or fully understood by a single individual. To complete these projects, a team of developers needs to work together to build individual components that fit together to build a complete system.

Large software projects are inherently risky. They take up huge chunks of the engineering organization's time and resources in an effort to build something that no one fully understands and that no one fully knows will succeed or not in the end.

---

75. https://en.wikipedia.org/wiki/Supply_chain_attack
76. https://www.fortinet.com/resources/cyberglossary/supply-chain-attacks

It's impossible to completely eliminate the risk involved in these large initiatives, but there is a useful tool for managing the complexity: decomposition.

Decomposition involves breaking down the problem into smaller and smaller pieces until each individual piece can be comprehended and completed individually. When decomposing large-scale projects, look for patterns or common components of work within the requirements and group them to create boundaries around related tasks. Doing this will help you expose natural hierarchies that simplify complex systems.

Breaking down tasks into smaller, more manageable chunks has the added benefit of exposing relationships and dependencies between the tasks. You may find that one task has to be completed before another can begin, or you may be able to identify tasks that can be worked in parallel by you and your team members so that the team can move quickly. Sometimes, things need to be built in a specific order, so use this technique to help expose critical dependencies and identify risks that may delay the project or prevent your team from meeting their deadline.

When to use decomposition:

- **When dealing with large projects:** In an agile development shop, you would break down long-term initiatives into medium-term epics, which are further broken down into short-term user stories. You can then help prioritize the order of user stories that should be worked on first.
- **When refactoring large pieces of a codebase:** Big changes equal big risk. Break up the changes into small pieces and refactor them piece by piece over time. There's much less risk in deploying incremental changes to a production environment than there is to deploying one large change.
- **When dealing with quarterly or annual goals:** You may have a few main short-term and long-term priorities, but what you need to do to meet those goals may not be obvious. Breaking them down to smaller subgoals will help you work backwards and figure out a plan of action.

No matter how large a task or project is, decomposing the problem is all about breaking down the requirements into smaller puzzle pieces. While this allows you to organize the pieces so they are easier to understand, what it really comes down to is managing risk by planning ahead.

10.3.2 PLANNING AHEAD

"Every battle is won before it's ever fought."

— Sun Tzu

*"Part of our strategy is getting the programmers to think
everything through before they go to the coding phase. Writing
the design documents is crucial, because a lot of simplification
comes when you see problems expressed as algorithms."*

— Bill Gates[77]

Whether you're assigned a ticket to work on or you're able to choose which ticket to pull in next, taking a little time to put a plan together can go a long way in reducing wasted time and effort on coding the wrong solution. Depending on how much information is in the ticket, it may be a straightforward change that's been thought through already, which is great! But there will be times where you don't have quite enough information from the ticket, and you'll need to do some research and planning before writing any code.

A common habit among junior programmers is that they'll begin writing code as soon as they pull a ticket from the backlog. They may not fully understand the problem or may not have a complete grasp on the codebase, so they start making small changes here and there to see if they can come up with a solution that works. While you might find a good solution now and then, there's a good chance some other programmer on your team had a different implementation in mind, and oftentimes, theirs might be better because they understand the problem or the codebase better.

Coding without a plan is a mistake that telegraphs your inexperience to your manager and the rest of your team. It often results in a lot of rework because you don't fully think through the problem and have to change direction before coming up with the final solution. You should be deliberate with most changes and not settle on the first solution that comes to mind, because there are oftentimes better ways to solve a problem.

An easy technique you can use to reduce the risk of rework is to plan out your work ahead of time. In fact, this is a technique you'll be using quite a bit as a professional programmer. You'll find that this will improve your decision-making skills because you can work through different sce-

---

77. Lammers, Susan. *Programmers At Work*. Tempus Books, 1989.

narios and eliminate ones that are insufficient or that would be difficult to maintain or extend in the future. Planning gives you an opportunity to iterate on your solution before writing code, rather than having to rewrite large chunks of code. After all, it's faster and cheaper to refactor an idea on paper than it is to refactor code that's already been written.

Rewriting code is expensive—it can cost hundreds or thousands of dollars because you didn't consider the consequences or side effects of a solution before implementing it. You may have to toss out code you spent all day writing because your coworker pointed out an edge case you didn't think about ahead of time. It's frustrating when you have to throw out work, especially after spending a long time working on the solution. Planning ahead hedges against the risk of having to toss out code.

Planning ahead also helps you see the bigger picture of the problem you're trying to solve, because it forces you to think about important decisions upfront when it's cheap to iterate to better solutions. Over time, the software industry has adopted different tools and frameworks for writing and structuring your planning. Design documents are the most common tool used for planning out the technical details of a software project.

A design document is a template or worksheet that helps you think about how your solution will meet a set of technical requirements. The technical requirements describe what the end result should be, and your design document describes how your solution will meet those requirements. A thorough design document should contain everything you and the other developers need to write the code to satisfy the project's requirements. Once you've spent the time thinking through the solution, the design document will serve as a guide for you and the other programmers throughout the life of the project.

Not only does a design document serve as your guideline, but it allows your teammates to evaluate and peer review your ideas before you spend valuable time and money implementing your solution. It's nearly impossible to know every side effect and dependency of the code you're writing, so the more kinks you can iron out in the design, the easier your code will come together when it's time to write it. Spending time compiling your ideas in a design document forces you to think through the architecture and how it will integrate with other parts of your codebase, as well as helping you gather valuable feedback before spending developer-hours implementing the wrong solution or even solving the wrong problem.

A primary purpose of using design documents is to come to a consensus on a solution before implementing it, which helps avoid costly disagreements in the future. By getting all parties on board with a solution, you can be sure that what you deliver is what was agreed upon. And if the stakeholders come to you and try to increase the scope of the project or change direction, you can point to the design document and show them what they agreed to at the beginning of the project.

It may feel like more work up front, but it's much cheaper to change the design of a solution during the planning period than it is to make an expensive change once the code has already been written.

𝒫 RESOURCES

- How to write a good software design doc[78] (medium.com/free-code-camp)
- How to Write a Software Design Document (SDD)[79] (nuclino.com)
- Learning to like design documents[80] (jvns.ca)

10.3.3 **CODE REVIEWS**

While this one may seem obvious to most people, the number of teams that ship code to production without a proper code review process is probably higher than you'd think. Under tight deadlines and stressful or even lax work environments, it is easy to skip the code review process altogether, and that introduces the risk you'll ship buggy code to production. It's especially common on small teams or on newer projects because things change so quickly as you're building out a minimum viable solution.

While it may allow you to ship code faster, skipping code reviews comes at the expense of code quality. Adding a second pair of eyes to peer review your code increases the quality of your work because your coworkers might catch bugs that you didn't even know existed. In addition to catching syntactic errors or nitpicking on coding standards, your coworkers may also catch potentially dangerous errors in the logic itself. Code you were confident worked one way, may work a completely different way

---

78. https://medium.com/free-code-camp/
    how-to-write-a-good-software-design-document-66fcf019569c
79. https://www.nuclino.com/articles/software-design-document
80. https://jvns.ca/blog/2016/06/03/learning-to-like-design-documents/

if there's a misplaced operator or parentheses. Your coworkers may also have more knowledge about a specific part of the codebase that you're changing and can help identify unforeseen circumstances with the changes you're proposing.

There's no doubt that code reviews can be frustrating. You may think you've done a good job coming up with a good solution to the problem, and you're probably proud of the code you've written, but your teammates may pick apart your code and ask for changes. They'll ask questions about why you built something the way you did and suggest edits that you may not think are correct.

It's easy to get defensive when it feels like you're being attacked, but it's important to remember that they're not criticizing you personally. You're all part of the same team and it's *everyone's* responsibility to ship quality code. Try to keep in mind that they're just trying to help you make improvements to your code.

Plus, there are plenty of benefits to code reviews that you may not realize, such as:

- When you review other people's code, it helps you learn the codebase.
- The codebase is constantly changing, so it also helps you stay up-to-date with the modifications being made.
- You'll be exposed to new techniques and patterns from the code that your coworkers write, and it will help you write better code.
- Your coworkers may offer advice on a better way of solving a problem, helping you learn and grow as an engineer.
- Requiring one or two pre-merge code review approvals adds checks and balances to reduce the risk of shipping buggy code.
- Having your code reviewed forces you to tie up any loose ends and make sure your code works and has been tested before submitting it for peer review. Just knowing that your coworkers will catch bugs means you'll spend extra effort making sure your code works properly.
- Code reviews give developers a chance to enforce consistency within the codebase, from patterns to naming conventions and syntax.
- Code reviews help catch critical mistakes that are often overlooked or misunderstood by the author.
- Your coworkers will help ensure your code meets the project requirements as well as your organization's coding standards.

- Your coworkers may find performance issues in your code and suggest ways to improve the efficiency of your algorithms.
- Likewise, your coworkers may find security issues in your code that could compromise your business's credibility or, worse, your customer's data.

The list above is by no means exhaustive, and there are many more benefits to the process of reviewing code before merging it into the main branch. While it can feel like a burden and extra overhead to some programmers just starting their career, the benefits outweigh the costs in the long run based on the number of issues that are caught during the development phase instead of allowing them to slip through to the staging and production environments.

Code reviews are all about managing and reducing the risk involved in shipping defective code. Just like authors, researchers, and students need to have their writing peer reviewed, so do programmers. We're not able to catch every mistake, especially when we're deep in the weeds trying to get our code to compile correctly. Having other developers double-check your work benefits everyone in the long run.

🔗 RESOURCES

- Why code reviews matter (and actually save time!)[81] (atlassian.com)

10.3.4 **STATIC CODE ANALYSIS**

Static code analysis is the act of analyzing a codebase without actually executing its code. The technique is gaining popularity among software organizations, and many teams are adopting tools to help standardize and find vulnerabilities within their code.

There is an entire industry dedicated to automating static code analysis so that you can focus on what you do best, building value for your customers. Some of the more advanced static code analysis tools will scan your software dependency graph for vulnerabilities and alert you to any libraries that you should upgrade and replace due to security issues. They often use proprietary or open-source databases, maintained by security researchers, to track known software vulnerabilities.

---

81. https://www.atlassian.com/agile/software-development/code-reviews

> ⟩ EXAMPLE

Here are a few examples of some great static code analysis tools:

- SonarCloud[82] helps you quantify code coverage and identify security vulnerabilities, duplicate code, and code smells.
- Snyk[83] helps you find and automatically fix security vulnerabilities in your code, open-source dependencies, and infrastructure code so you can focus on building.
- GitHub's Dependabot[84] helps you keep your dependencies up-to-date by automatically opening pull requests against your GitHub repositories to install updates.

If your team doesn't already use static code analysis to aid in finding and fixing vulnerabilities, consider suggesting that they try out some tools. You'd be surprised at what vulnerabilities may be lurking in your codebase, and you can leverage these tools to harden your systems and build more reliable software.

> 🔗 RESOURCES

- Static program analysis[85] (wikipedia.org)

10.3.5 **AUTOMATED TESTS**

In the previous section[§9] you learned how an automated test suite can provide immense value to your team. Automated testing is so important that it's worth mentioning again, because it doubles as a way to manage and reduce the risk of introducing defects when making changes to existing code. A team with sufficient automated test coverage across their codebase can proactively catch bugs faster and cheaper before their code changes hit production.

Building good habits like writing unit and functional tests when you commit new code is one of the best things you can do as a junior programmer. If your team doesn't already have a test suite or a continuous integration system in place, use that as an opportunity to suggest one and implement it yourself. It's a lot of work up front, but it's a long-term invest-

82. https://sonarcloud.io/
83. https://snyk.io/
84. https://github.com/dependabot
85. https://en.wikipedia.org/wiki/Static_program_analysis

ment that will bring improvements to developer productivity for years to come.

> EXAMPLE

Here are examples of how automated testing can help you and your team:

- Automated tests lead to increased productivity, because you can make changes to parts of the codebase with confidence that you're not breaking existing functionality.
- Faster feedback loops because you can run the tests locally or on your continuous integration server as you're making changes. There's no need to deploy your code to hosted environments to make sure it's working properly.
- The overall software development life cycle can be shortened because you can make changes and write new tests to ensure the code is working properly.
- You're able to reduce the risk of introducing new defects because you can write test code that checks for specific edge cases and then run those tests over and over again.
- Automated tests allow you to focus on feature development and building for scale, rather than tracking down and fixing bugs introduced into the system when you make changes to legacy code.

If your team already has a continuous integration system in place, that's great. All you have to do then is build the habit of adding new tests with every change you make. You'll be surprised at how quickly your test suite grows, and pretty soon you'll have good coverage over the business-critical components of your system. The more test cases you can cover, the lower the probability of introducing regression issues into your codebase. And lowering the probability of introducing breaking changes lowers the risk when refactoring or making changes to the system.

RESOURCES

- Reducing Risk And Improving Agility With Test Automation[86] (radiant.digital)

86. https://radiant.digital/reducing-risk-and-improving-agility-with-test-automation/

- Test Automation: The Ultimate Guide[87] (leapwork.com)

10.4 *Postmortems*

> *"Failure is only the opportunity more intelligently to begin again."*
>
> — Henry Ford[88]

This whole section has been about managing and reducing risk, but an unfortunate fact of life is that it's nearly impossible to completely eliminate all risk involved in writing software. With any moderately complex software, things will go wrong at some point. And sometimes things will go *very* wrong. Failure is inevitable, and at some point, you'll be pulled into an incident. When these incidents happen, it's important to use them as learning experiences and take the time to reflect on the preceding events in order to better understand how and why they happened. In doing so, you'll be able to learn from your mistakes and make any appropriate changes to prevent them from happening again in the future.

The best thing you can do in the aftermath of an incident is to capture and document what happened leading up to, during, and after the incident so that you can reflect, learn, and share that knowledge with others within your organization. This process is known as a postmortem.

An incident postmortem should bring people together to discuss and document the details of an incident:

- What was the timeline of events leading up to and during the incident?
- What was the ultimate root cause?
- What was the impact on the customers and the organization?
- What actions were taken to mitigate the failures and get the system back to a stable condition?
- What steps, if any, should be taken to prevent the same thing from happening again?

If you and your team are able to set aside time to put together a root-cause analysis after a major operational incident, then you're setting yourself up for the opportunity to improve yourself, your teammates, and your

---

87. https://www.leapwork.com/test-automation
88. Ford, Henry. *My Life and Work*. Doubleday, 1923.

team's software development processes. When you learn from your mistakes, you're able to reduce the risk of making those same mistakes in the future, but it takes time and effort to assess the impact and damage after the dust has settled. A postmortem is a useful framework for sharing knowledge and learning from incidents. Its ultimate purpose is to help organizations turn negative events into forward progress.

Postmortems can be difficult, however, especially if one highlights a mistake or oversight you personally made. You or one of your colleagues may be embarrassed or nervous to share details within your organization. Successful postmortems should be blameless and focus on finding a solution to prevent the root cause from happening again, not on pointing fingers and assigning criticism.

Your goal should be to bring people together in a constructive and collaborative environment that allows everyone to contribute to the progress and evolution of the organization. Postmortems are designed to build trust among team members, across teams, and even with customers. Some companies choose to publish their postmortems publicly in order to show their customers transparency and rebuild confidence in their products.

> ℰ RESOURCES
>
> - The importance of an incident postmortem process[89] (atlassian.com)
> - Making the Most of a Project Post-Mortem[90] (builtin.com)
> - How to run a great software incident post-mortem[91] (lead-dev.com)

You don't need to wait for an incident to reflect and learn from your past, however. There's another framework, called retrospectives, commonly used by many modern software companies.

10.5 *Retrospectives*

While progress may seem linear from the outside, behind the curtains it is sometimes a chaotic and sloppy process to get where you're trying to go.

---

89. https://www.atlassian.com/incident-management/postmortem
90. https://builtin.com/software-engineering-perspectives/project-post-mortem
91. https://leaddev.com/reporting-metrics/how-run-great-software-incident-post-mortem

Things never go according to plan *all* the time, and you need to learn to adapt to changing requirements and external influences.

As professional software developers, it is our job to master the art and science of delivering quality software on time and within the project requirements. To do this, we often reflect on our current processes and continually improve the way we deliver software. This act of continuous reflection and improvement is enabled by a framework called a retrospective.

The idea behind the retrospective was originally published in 2001 as the twelfth and last bullet point of the Agile Manifesto,[92] which states that:

> *"At regular intervals, the team reflects on how to become more effective, then tunes and adjusts its behavior accordingly."*

Retrospectives are meant to offer a framework for your team to evaluate itself and devise a plan to address any areas of improvement for the future. By reviewing and analyzing our past projects, we can determine which processes worked well and identify where we can improve the ones that broke down. This allows teams to add, modify, or remove processes in order to become more productive on the next development cycle.

Retrospectives are designed to involve the whole team and to encourage everyone to be honest and offer insights and opinions on what went wrong and how it can be improved. When teams identify areas of improvement and take action to improve going forward, they're taking proactive measures to reduce the different types of risk we discussed at the beginning of this section.

Remember, the ultimate goal is to manage and, if possible, eliminate risks that would prevent you and your team from delivering high-quality software on time and on budget.

🔗 RESOURCES

- Why run a retrospective?[93] (atlassian.com)
- The Importance of the Retrospective[94] (dan-schreiber.com)

---

92. https://agilemanifesto.org/principles.html

93. https://www.atlassian.com/agile/scrum/retrospectives

94. https://dan-schreiber.com/blog/2015/2/9/the-importance-of-the-retrospective

11 How to Deliver Better Results

We all want to write great code and feel like we're contributing to the success of our team, but it takes more than just writing clean code or finding the perfect abstraction. Even as an individual contributor, there will be things you need to manage, such as your time and productivity. You're directly responsible for making sure you're using your time wisely and keeping your output high, but that's easier said than done. Some days, you may feel like you're getting a lot of work completed, while other days, you'll feel completely stuck and not sure what to do next.

Delivering results is all about finding your personal groove that'll allow you to churn through tasks and ship some actual code on a regular basis. That doesn't mean you should lose sight of producing quality work, however. Your first focus should always be on *quality* code. If the code isn't up to your team's standards, then you should absolutely spend additional time cleaning it up so it's ready for production. There's no point in moving quickly if you're shipping half-finished code that's full of bugs—you'll just be shifting the burden on to the rest of your team to find, fix, and maintain the defects in your work.

To be a productive software engineer, you should strive to continuously move forward and make progress toward building value and managing the risks involved in shipping code. So, let's dive in and look at what you can do to increase your productivity and deliver better results.

11.1 *Perfect Software Doesn't Exist*

The first point to remember is that perfect software doesn't exist. There will always be multiple ways to solve a programming problem, and each comes with its own trade-offs. Some may be faster or scale more efficiently, while others may be easier to read and understand, maintain, or extend. Rarely will there be a perfect solution to a problem because there will always be compromises that come with each option.

There's a model of constraints called the Project Management Triangle that states that the quality of work for a given project is bound by three things: the budget, the deadline, and the scope of a project. While it's possible for the project manager to trade between these constraints, they're only able to optimize for two of the three. Programmers and engineering managers commonly refer to this model when they state that software can be good, fast, and cheap, but the catch is that you can only choose two.

- Fast and cheap projects tend to produce lower-quality software
- Fast and high-quality software will be expensive to produce
- High-quality and cheap software will be slow to produce

Figure: Project Management Triangle.

There will be times when you'll be forced to make trade-offs or cut corners in order to meet a deadline or keep a project under budget. It happens all the time in software development. The reality is that you have to accept that some code you ship to production may be sloppy, inefficient, or could be optimized better if you had more time.

Junior engineers tend to keep tweaking and fine-tuning a solution because they want their code to be clean and perfect, while senior engineers tend to have a better feel for when a piece of code is "good enough" to ship, and then they move on to the next task. If you look closely when reviewing pull requests from your coworkers, you'll begin to notice code written by senior engineers often has a hint of inelegance to it. They understand that it's not worth the time to optimize and fine-tune each and every algorithm, abstraction, or variable name.

Knowing when to ship "good enough" code is a skill that you'll build over time, not something you can learn overnight. The more software you ship, the more you'll understand where to focus your time and effort, and where you can write some quick and dirty code that gets the job done.

11.2 *Find Tools That Work for You*

> *"If you ever talk to a great programmer, you'll find he knows his tools like an artist knows his paintbrushes."*
>
> — Bill Gates[95]

As programmers, we have thousands of tools available that help us perform our jobs, and choosing the right tool for the job can increase your productivity tenfold. But a lot of powerful tools go underutilized because the programmer doesn't understand how to use them to their full potential.

Before you can truly be productive as a programmer, you need to develop a deep understanding of the tools at your disposal. You should strive to regularly add new tools to your toolbox, but with each new tool you add, be sure to take the time to learn its advantages and disadvantages so you can know when (and when not) to reach for it. No tool is perfect for all situations, so don't fall victim to relying on just one tool because you're comfortable with it. Every tool has its strengths and weaknesses, and a good programmer knows when it's best to use each one. These are the tools with which you will build great things, but you can't do that until you know them inside and out.

As programmers, we commonly compare writing code to crafting software. Almost every line of code is handwritten for a specific problem you're solving, so it feels natural to consider programming as a craft. There are quite a few similarities between programmers and other types of craftsmen like plumbers, electricians, and carpenters. Craftsmen are able

---

95. Lammers, Susan. *Programmers At Work*. Tempus Books, 1989.

to look at a problem, devise a solution, and use their tools to solve the problem. Sounds a lot like programming, right?

If you ever step into a carpenter's workshop, you'll immediately notice a myriad of tools they've amassed over the course of their career. You'll find some general-purpose tools such as saws, routers, and drills that can be used for a variety of things. But you'll also find very specific tools meant to do one thing and one thing very well, such as planers, clamps, custom-built jigs, and hand tools to gradually shape and refine the surface of a workpiece. Each tool in a carpenter's workshop has a specific purpose to solve a specific problem, and an experienced carpenter knows exactly when each is the best tool to accomplish the task at hand.

Additionally, woodworking is extremely dangerous when you're working with power tools and razor-sharp blades. To perform their job safely, an experienced carpenter also needs to know the limitations of their own skills and of each tool. If they use a tool incorrectly or are not paying attention, they could lose a limb or possibly their life.

While programming certainly isn't as dangerous as carpentry or other professions, we too need to know the strengths, weaknesses, and limitations of our tools if we want to do our jobs efficiently and deliver quality software. Thanks to the explosion of open-source software available to us, choosing the right tool for the job is much harder today than it was ten or even twenty years ago. Not only are there thousands of programming languages, frameworks, and libraries to choose from now, their quality is continuously increasing as the industry learns how to build better software.

Modern software development, especially full-stack web development, requires programmers to switch between dozens of languages, frameworks, and libraries across all parts of the software stack. It's nearly impossible to master all of the tools you work with on a day-to-day basis, so don't feel pressure to learn every feature, syntax, and pattern. At a minimum though, you should at least have a good understanding of your tools and *why* you use one over another.

So, now let's look at different types of tools and what you should know about them.

11.2.1 IDES

The integrated development environment (IDE) you choose for writing code can have a bigger impact on your productivity than you may realize. Modern IDEs are designed specifically to maximize developer productivity and are incredibly powerful. Your IDE should be considered one of the most important tools in your toolbox.

If you're only using your IDE as a text editor to write and edit code, you're likely missing out on powerful features you may not be aware of. Not using the full power of your IDE would be similar to buying an expensive chef's knife and only using it to spread butter on your toast. That's not what a chef's knife is made for, and you wouldn't be using the knife to its full potential. A good IDE used to its full potential can feel like it's adding superpowers to your workflow and productivity because it can do a number of things that are difficult for humans.

> ❯ EXAMPLE

So, how can an IDE make you more productive?

- Save time by using shortcuts to switch between files quickly without leaving the keyboard.
- Save time with shortcuts to open files from paths that were copied from stack traces, logs, documentation, etc.
- Save time when needing to look up standard library functions with built-in API references.
- Save time when refactoring by using smart find and replace features to search for and rename variables, class names, and file names in individual files or across your entire codebase.
- Use code completion to reduce errors and prevent typos when typing variables and class names.
- Use the power of debuggers to step through your code to track down bugs and performance issues in your code.
- Catch syntax errors early with syntax highlighting; there's no need to refresh the browser or wait for your code to compile.
- Extend the IDE with plugins that offer new functionality, or build your own tools directly into the IDE.

If you haven't already, I'd encourage you to try a few different IDEs to test drive them and see which one you like the most. Every programmer has their own unique workflow, so I'd recommend you stick with whatever IDE you feel most comfortable working with. After all, you'll be spending a significant part of your career writing code in an IDE, so it's worth it to spend the time to find one that works for you and learn how to use it to its full potential.

> ❯ EXAMPLE

Some IDEs to consider trying out:

- IntelliJ IDEA.[96] The flagship IDE created by JetBrains. Although it's mainly geared towards Java development, JetBrains also offers the same IDE specialized for other applications such as Ruby, Python, Golang, PHP, .NET, and JavaScript/TypeScript.
- Android Studio.[97] An IDE created by JetBrains that is geared specifically toward Android development.
- Visual Studio Code.[98] Created by Microsoft, this has become one of the most popular IDEs used by developers due to its large ecosystem of extensions to add additional functionality and customization.
- XCode.[99] Created by Apple, this IDE combines tools to design, code, test, and debug iOS and MacOS applications in one application.

11.2.2 THE COMMAND LINE

Integrated development environments typically include GUIs, which make them user-friendly and easy to get started with. We're all familiar with browsers and desktop applications, so a GUI-based code editor seems like a natural first step when it comes to writing and editing code. But there are a lot of things that you cannot do with GUI applications.

Command-line terminals, on the other hand, often seem confusing and intimidating to those who are unfamiliar with them. You'll probably have some difficulties when you're first learning how to navigate and run commands on the command line, but over time, these actions will become second nature and easier to remember.

The command-line terminal will become an essential tool on your road to becoming a senior engineer. It's probably the most powerful tool in your toolbox, even more so than your IDE, so it's a good idea to get familiar with it as soon as you can. The command line can amplify your skills and increase your productivity by orders of magnitude, but you need to know how to use it to its full capabilities before you can unlock that potential.

---

96. https://www.jetbrains.com/idea/
97. https://developer.android.com/studio
98. https://code.visualstudio.com/
99. https://developer.apple.com/xcode/

>> EXAMPLE

Here are some examples of how the command line can make you more productive:

- Save time by creating, copying, renaming, and deleting files without leaving the keyboard.
- Quickly search for phrases and patterns across your entire codebase or file system.
- Chain commands together to pipe output from one program to another, allowing you to create powerful pipelines to process data.
- Create your own shortcuts and aliases to speed up repetitive tasks and commonly used commands.
- Create your own utilities to extend the power of the command line.
- Write and execute scripts to automate larger tasks.
- Install dependencies, frameworks, libraries, and other tools with a few commands.
- Monitor and analyze current running processes and system diagnostic information on your host machine.
- Access remote servers and perform network operations to monitor and analyze network traffic.

There are endless things you can do with the command line—you're only limited by your imagination. The more you use it and become comfortable with never leaving the keyboard, the more creative you'll get when it comes to thinking up new ways to solve problems directly from the command line.

Don't worry about speed when you're first starting out. We can't all be as smooth as hackers in the movies are, and the good thing is we don't need to be. Just focus on learning the command line at your own pace. The speed will come with repetition and practice. The more important thing to focus on is getting comfortable and learning how to solve the problem you're working on with the commands you have at your disposal. Everything else will come naturally, and you'll be boosting your productivity in no time.

- The Art of Command Line[100]

11.2.3 PROGRAMMING LANGUAGES

It goes without saying that if you're hoping to make a career as a professional programmer, you should strive to have an expert-level knowledge of at least one programming language.

It might sound silly because it's so obvious, but there are a lot of developers who jump from language to language because they're chasing the hot new trendy tool that everyone's talking about. If you're serious about having a long and successful career as a software engineer, you need discipline and focus to stick with one language long enough to become an expert in it.

Ideally, the language you will pick will pay the bills by enabling you to leverage your expertise into lucrative employment opportunities. Spending the time to become an expert in one language pays off because it goes a long way toward providing job security.

Whatever language you choose early in your career, stick with it until you become an expert in it. Learn the language's strengths—what problems is it really good at solving? What are its weaknesses? It's also important to learn its inconsistencies and any pain points so you know when to avoid them.

Only after you feel comfortable writing code at an advanced level in one language should you begin to branch out and learn other languages. While you'll want to stick with one or two languages in the first few years of your career, learning how to write programs in different languages is an important skill to learn as you gain more experience. You'll learn new patterns, techniques, and even new programming paradigms by learning new languages. It will force you to become a better programmer as you apply what you've learned to the code you write, even if it's in a completely different language. Solving problems in different programming languages forces you to rely on your knowledge of the foundational concepts of programming rather than the syntax of one language over another.

At some point, you'll work on a new project and have the opportunity to decide which language to choose. If you have the knowledge about the strengths and weaknesses of different languages, you'll be able to make

---

100. https://github.com/jlevy/the-art-of-command-line

better decisions on which to choose. And in some cases, choosing the right language can increase your team's productivity by an order of magnitude.

🔗 RESOURCES

- What programming languages should a modern day programmer have in his/her arsenal?[101] (quora.com)
- Top 43 Programming Languages: When and How to Use Them[102] (raygun.com)

11.2.4 UBIQUITOUS TOOLS IN THE INDUSTRY

Knowledge of the right tools at the right times can unlock amazing opportunities for you and your career.

Some tools and technologies are so common across the software industry that all software engineers should at least have a general understanding of them. Familiarity with these technologies make you a well-rounded engineer.

Tools that every software engineer should be familiar with:

- **Linux**

  - An interactive course to learn the basics of Linux[103] (linuxjourney.com)
  - Introduction to Linux (LFS101x)[104] (linuxfoundation.org)

- **Git**

  - Git Documentation[105] (git-scm.com)
  - Become a Pro at Git with this Guide[106] (atlassian.com)
  - Git - the simple guide[107] (rogerdudler.github.io)

- **Docker**

  - Docker overview[108] (docker.com)

101. https://www.quora.com/What-programming-languages-should-a-modern-day-programmer-have-in-his-her-arsenal/answer/Joshua-Levy
102. https://raygun.com/blog/programming-languages/
103. https://linuxjourney.com/
104. https://training.linuxfoundation.org/training/introduction-to-linux/
105. https://git-scm.com/doc
106. https://www.atlassian.com/git/tutorials/what-is-git
107. https://rogerdudler.github.io/git-guide/

- A Docker Tutorial for Beginners[109] (docker-curriculum.com)

- **SQL**

  - Intro to SQL: Querying and managing data[110] (khanacademy.org)
  - Learn SQL with simple, interactive exercises[111] (sqlbolt.com)

- **HTTP**

  - An introduction to HTTP: everything you need to know[112] (freecodecamp.org)
  - HTTP Reference[113] (developer.mozilla.org)

- **Makefiles**

  - Learn Makefiles with the tastiest examples[114] (makefiletutorial.com)
  - GNU make manual[115] (gnu.org)

This is just a small list of common tools that are universally used by software developers at some point in their career. You don't need to become an expert in each tool, but having at least a general understanding of each of them will round out your technical skills.

> 🔗 **RESOURCES**
>
> - Programmer, know thy tools![116] (dzone.com)
> - Roadmap.sh[117]

---

108. https://docs.docker.com/get-started/overview/

109. https://docker-curriculum.com/

110. https://www.khanacademy.org/computing/computer-programming/sql

111. https://sqlbolt.com/

112. https://www.freecodecamp.org/news/http-and-everything-you-need-to-know-about-it/

113. https://developer.mozilla.org/en-US/docs/Web/HTTP

114. https://makefiletutorial.com/

115. https://www.gnu.org/software/make/manual/html_node/index.html#SEC_Contents

116. https://dzone.com/articles/programmer-know-thy-tools

117. https://roadmap.sh/

11.3 *Find a Process That Works for You*

Once you've mastered your tools, the next thing to focus on is your own process for producing software. Every programmer approaches software development differently, and what works for some people may not work for others. There are several development methodologies and ways to solve problems, but I'm going to share a process with you that I've found works for a lot of programmers. It's simple and straightforward, and it helps you stay focused on the important thing—delivering working software. So, here's the process:

> *"Make it work, make it right, make it fast."*

That's a quote from Kent Beck, the creator of extreme programming and one of the original signatories of the Agile Manifesto. Kent has shaped programming in many ways, and this technique will hopefully shape the way you approach programming. Following this simple pattern will help you manage the complexity of your own solutions and prevent you from trying to do too much all at once.

Let's dive into it a little more.

11.3.1 MAKE IT WORK

When you first set out to write new code for a feature or bug fix, you should be fully focused on proving *the problem can be solved*. Without this step, nothing else matters, so you should always value a working solution, even if it's messy, over something clever that doesn't compile.

◇ IMPORTANT You are allowed to violate principles of good software design in this phase because your only goal is to make things work and work repeatedly.

This phase of software development can be compared to writing the first draft of an essay, so it's okay to have ugly and messy code while you're working towards a solution. You don't need to come up with good variable names or a good object-oriented design, and your code doesn't need to be elegant. It's completely fine to start out scripting a solution at first, if that makes it easier for you to focus on solving the problem.

Once you are able to compile your code and run it, and you have somewhat of a working solution to the problem you're solving, save your progress. Commit your changes to version control. Doing this will give you a good stopping point where you know you have a working solution. Now that you have at least *something* working, you can begin to refactor and improve upon the solution.

11.3.2 MAKE IT RIGHT

Once you have a solution that you've proven works repeatedly, it's time to move on to the next phase, which is to make it right.

To follow the essay comparison, this would be the revision phase. Here you should focus on improving the design, reliability, and readability of your code. This is where you refactor the code you just wrote—clean up your naming conventions, introduce abstractions and interfaces to aid in extensibility, and add tests. Make sure to cover all your edge cases and remove any hard-coded values you may have used in the first phase.

Your goal is to clean up the code so that when you return to it in the future you'll be able to understand what it's doing and easily change it if needed. There may be other programmers on your team that could be extending it in the future, so keep that audience in mind when refactoring and documenting.

Your code should be bulletproof at the end of this phase. You should be confident in your solution and comfortable shipping this code to production.

11.3.3 MAKE IT FAST

> *"The real problem is that programmers have spent far too much time worrying about efficiency in the wrong places and at the wrong times; premature optimization is the root of all evil (or at least most of it) in programming."*
> — Donald Knuth, *The Art of Computer Programming*

This last step, which happens to be the hardest, is to improve the performance of your solution by making it fast. The reason it's the hardest step isn't necessarily because performance gains can be hard to come by, but because it can often be hard to find the time or resources in order to properly improve the performance of your solution.

After you've finished the "make it right" step, your code should technically be ready for production. You should be able to ship it and move on to the next feature or bug fix that provides value for your customers. And this is what makes the "make it fast" step so difficult, because you may be able to provide more overall value by ignoring this step completely and working on a new ticket.

In some cases, shaving a few milliseconds off the run time for your solution may not be the best use of your time, because that's not what's adding the most value for your customer. It's a good idea to communicate with your manager before starting this step, because they may have other projects that take priority over optimizing your solution.

In the end, your goal is to provide software solutions that provide value to your customers, so it's important to be aware of how much time you're spending on performance optimizations. Most of the time, "good enough" will be fine to satisfy the requirements so you can move on to the next project coming down the pipeline.

11.4 *Solve the Problem First*

In the earlier section[§10] about managing risk, we explored the importance of planning ahead when working on large projects. That idea also applies to smaller, individual tasks you work on as well. When you pull a new ticket to work on, what you *should not do* is start coding right away, even if you think you may know how to solve the problem. It's a trap a lot of developers fall into, and it's potentially a risk because you could be wasting your time and effort implementing the wrong solution.

If you don't have a good understanding of the problem, the requirements, and the acceptance criteria, you run the risk of shipping code that doesn't actually solve the issue or that may not be the optimal solution.

So, what should you do when you start a new task?

1. Read the task description and take note of any requirements. It's sometimes helpful to make a checklist of requirements so you can be sure you've checked them all off before submitting your code for review.
2. Read the task description again to make sure you fully understand the problem and what needs to be done. Sometimes it takes a few passes to fully understand what is being requested in the ticket.
3. Write down any questions you might have about any of the requirements in the ticket. It helps to get your thoughts out of your head and written down in case you get distracted or need to temporarily shift your focus to another task.

4. Browse through the codebase to get familiar with the code you think you'll need to change. There may be technical debt you'll need to work around, or there may be some abstractions that your solution may need to fit into, so it's a good idea to poke around the codebase to make sure you know what you're working with.
5. Come up with an implementation plan. By now, you should have a good understanding of the problem and any technical limitations you may need to work around.
6. Start coding.

Once you've worked through all the steps above, the actual coding part should hopefully go smoothly. If you've done good research and planned out what changes need to happen before writing any code, it should be fairly straightforward to implement your plan. Most of the work was already done ahead of time, so you just need to translate that plan to code a computer can understand.

By doing all the heavy lifting during the planning phase, you're reducing the probability that you'll need to change direction and rewrite code. Don't assume that code is easy to change and that you can just delete it and write it again. Changing direction after you've begun implementing a solution costs money, especially when you're dealing with large codebases with millions of lines of code.

◇ IMPORTANT   It is faster and cheaper to refactor an idea than it is to refactor code.

Solving the problem first allows you to move faster because you've already identified how your solution will fit into the existing code, and you already have an idea of what code needs to be refactored, improved, or moved around to implement your solution. Plus, you can present your solution to your teammates to gather feedback before writing any code. It's possible they may come back with questions or concerns, or they may suggest a better approach that is more efficient or simpler than your original idea. (Occasionally, on complex and important problems, an engineer might both design and prototype multiple solutions—for example, to then compare their performance. But this is a rare situation.)

In order to move quickly and deliver results consistently, you need to learn how to think strategically. Spending time upfront planning your implementation before coding it will reduce costs and time later when it comes to coding your solution.

🔗 RESOURCES

- How to Solve any Programming Problem[118] (medium.com/before-semicolon)
- Problem Solving[119] (denvaar.github.io)
- Start with pen and paper[120] (sethetter.com via Internet Archive)

11.5 *Take Ownership*

If you're working towards a promotion to a senior role, your manager may tell you that they'd like you to "take more ownership" of certain projects or tasks that your team is responsible for. To many people, "taking ownership" sounds vague and ambiguous the first time they hear it. Take ownership of *what*, exactly?

There are a lot of differing opinions on what taking ownership means, so if your manager encourages you to do so, the best thing to do is to simply ask them what taking ownership means to them. It's always good to clarify their expectations to make sure you don't miss something they are expecting you to do.

A developer taking ownership of something commonly means they are taking on more accountability for the success of a certain project or task. They are not necessarily the only one responsible for the success or failure of a project, but they will have a bigger influence on the outcome.

》 EXAMPLE

A junior engineer will typically work on tasks as part of a larger project. The requirements for those tasks were probably defined by someone more senior than them, either a senior engineer or a

---

118. https://medium.com/before-semicolon/
     how-to-solve-any-programming-problem-44883180c730
119. https://denvaar.github.io/posts/problem_solving_example.html
120. https://web.archive.org/web/20230430170508/https://sethetter.com/posts/
     start-with-pen-and-paper/

manager, and the junior engineer is just implementing a solution to
meet those requirements.

As that junior engineer takes on more ownership, they'll start to
get more involved in the planning aspects of the projects rather
than just implementing someone else's plan.

That junior engineer will now add more tickets to the backlog and
make sure all the project requirements are defined so that they can
be sure the project is meeting the customer's needs. They'll help
research, plan, and prioritize in case there are any internal or external dependencies that dictate when things need to be done. And
they'll coordinate the release of the code to production, measure
its success, and contribute to the ongoing maintenance of the project they led from start to finish.

That junior engineer took on more responsibility for the outcome
of the project.

The path to success for a project is never as simple as it looks, and the
more ownership you take on as you work towards a promotion, the more
ambiguity you'll be dealing with, which can be difficult. Not only do you
need to chip away at tasks to get to the finish line, but you also need to
look at the end goals of the project and work backwards to figure out a plan
for how to reach the finish line. You may not even know where to begin at
times, but that's part of taking ownership and growing into a senior role.
Taking ownership means figuring out a path forward, even when faced
with uncertainty about how to proceed.

Ownership isn't always directly tied to specific projects, either. Sometimes taking ownership of areas involved in a team's development process
is a good way to show maturity. When a developer's mindset shifts from
advancing their own technical abilities to advancing those of their team,
they begin the process of thinking like a senior engineer.

While their main projects and priorities always come first, senior
developers don't sit back and wait to be told what to work on. They
actively seek out areas where they can improve the codebase or, even better, improve their team's development process.

Senior engineers learn to put their team's needs above their own. That
often means that senior engineers work on things such as:

- Helping unblock other developers if they are stuck or if they are waiting on a code review.

- Working on the mundane but important tasks, such as bug fixes or writing documentation.
- Helping to teach other engineers and transfer knowledge across the team.

In essence, taking ownership is all about taking on more responsibility for the outcome and results of individual projects and for the results for your team. As you work towards a senior title, try to identify areas where you can contribute more on individual projects, and try to take on more responsibility in areas where you think you can improve your team's over-all output. The more results you can help deliver, whether through your own work or enabling others, the higher the chances of earning that promotion to a senior role.

🔗 RESOURCES

- Taking Ownership Is The Most Effective Way to Get What You Want[121] (effectiveengineer.com)
- Getting Things Done When You're Only a Grunt[122] (joelonsoftware.com)
- Ownership — the secret ingredient[123] (medium.com/sears-israel)

12 How to Communicate More Effectively

It's a popular stereotype that programmers are often introverted, reclusive, and lacking in the social skills department. Even though this doesn't accurately reflect the industry as a whole, it's been perpetuated in anecdotal stories of old-school hackers in the early days of Microsoft, Apple, Facebook, and many more tech giants. There's a common theme in Hollywood movies in which the genius coder hacks into the mainframe from his dark basement, surrounded by empty pizza boxes and energy drinks. But while the act of programming has always been between a human and a machine, the software products used by people throughout the world are a result of

---

121. https://www.effectiveengineer.com/blog/take-ownership-of-your-goals
122. https://www.joelonsoftware.com/2001/12/25/
     getting-things-done-when-youre-only-a-grunt/
123. https://medium.com/sears-israel/ownership-the-secret-ingredient-bf01e2b50acb

collaboration and communication between many different people. Shipping software at scale is a team effort, and it takes good communication skills to deliver quality products.

Over the years, we've gotten better at building software. We've learned from our mistakes; invented new algorithms, software patterns, user interfaces, and development methodologies; and built languages and tools that have increased the speed and quality with which we can produce software. As the industry changes, so has the role of programmers. It's no longer enough to be technically competent—communication skills are just as important as technical ability and will continue to play a critical role in software production for years to come.

Although the act of programming is mostly an individual one, working on a team with other technical and nontechnical people will be one of the hardest things you do in your career. Human nature is incredibly complex and unpredictable at times, and as programmers, we're required to interact with many people throughout the software development process in order to perform our jobs.

It may be uncomfortable at times, especially for junior programmers because you're learning how to write code at an advanced level while simultaneously learning how to interact with other programmers and business stakeholders. There will be times when you'll struggle to communicate your ideas clearly, and it'll be frustrating. And there will be times where you'll be intimidated because you can't get your thoughts out while all eyes are on you. We deal with abstract thinking quite often in our industry, and it's sometimes hard to convey those ideas to others in a way that helps them understand your exact point of view. Being able to convey technical ideas clearly and concisely takes good communication skills, and being able to communicate clearly is a skill that can be learned over time, but it requires practice.

Technical knowledge is relatively easy to acquire nowadays, especially with the internet as a distribution channel and the number of books, videos, and other resources we have at our disposal. The ability to communicate and connect with people on a personal level is much harder and may even require introspection and difficult changes to your own personality in order to improve your skills. Unlike a new programming language or design pattern, good communication skills are not something you read about once and understand right away. These skills require years of work and putting them into practice in your day-to-day conversations before they become second nature.

To be successful in modern-day software development, programmers need to be well-rounded individuals who can communicate cross-functionally with other teams. You might be asked to collaborate with Marketing, Sales, Customer Success, and other teams within the company to produce better software for both internal and external users. Some developers are even tasked with communicating directly with customers and clients. Occasionally, you may need to work with programmers at other companies or on other teams to build an integration between your system and theirs. The other programmers may have their own priorities (and their own technical limitations) for their part of the integration.

Good ideas themselves are no longer enough to build great software products. You need the ability to communicate your ideas so that they are properly received by your audience and the entire team has bought into your ideas. And finally, as we'll see in a bit, good communication is not just about talking, but also about listening to others.

Most senior developers understand that their communication skills are what sets them apart from their peers, and their skills will position them as a trusted source of knowledge within their organization. Good leaders are often the best communicators, and programmers need good communication skills if they aspire to lead projects and teams.

If you want to be recognized for your work, influence the direction of your team and your products, or earn a promotion, you need to learn how to interact well with other people and build meaningful relationships along the way. Great communication skills are a differentiator among programmers and will help you stand out amidst your peers of similar experience. What many programmers fail to understand is that the best communicators have a special skill that surpasses pure technical knowledge.

As I mentioned earlier, this skill can be learned over time, and anyone, including you, can master it. So, what can you do right now to improve your communication skills? Let's take a look at a few ideas.

12.1 *The Golden Rule*

Words are powerful, and you can hurt people deeply with just words alone. All it takes is a few select words to ruin professional and personal relationships that you've worked for years to build. It may sound a bit silly that "be good to others" is advice that will help you in your career, but when smart people with lots of passion work closely with one another, it's easy to forget. Difficult decisions get made every day, at both the business and technical levels, and sometimes people come out on the losing end of those decisions. Sometimes your ideas won't be chosen, or someday you may need to make decisions that affect people's careers.

Words have consequences, and it's important to choose the right words and do your best not to hurt someone's feelings. Words can create harm to other people, and they can create conflict with your coworkers.

One of the most important things to remember is to be aware of your emotional state when communicating with others. You may be pulled into an incident that wasn't your fault, or you may be participating in a heated discussion about the best design for a new system architecture. Regardless of what it is, you should always act and speak with empathy and professionalism. Keeping your composure when tensions are high is not something many people will notice, but losing your cool under pressure is certainly something that *everyone* will notice.

It's not just about staying calm under pressure, either. It's also about the tone you use in your conversations and that you use with other people, regardless of their seniority or job function. All of these variables affect how you should convey your thoughts if you want to get your point across and position yourself so people will take your ideas seriously. It starts with yourself, and that may be hard for some people to grasp. The sooner in your career that you focus on building rapport with your coworkers, clients, and customers, the easier it will be to gain support for your ideas. And it starts with awareness of who you're talking to and how you're talking to them.

12.2 *Know Your Audience*

Understanding the audience you're communicating with is an important principle to keep in mind for effective communication. Knowing who you're speaking with and their level of understanding about a topic will often dictate how the conversation will play out. Are you communicating with your boss, another programmer, a nontechnical coworker in another department, or an external client?

Depending on how technical your audience is, you may need to change how you explain certain topics. You might be able to discuss the details about API schemas, HTTP status codes, and how CORS requests should be handled between your backend and frontend applications with your fellow programmers, but a customer success representative or a marketing manager may have no idea what those topics mean. Sometimes it can be difficult to explain technical topics to nontechnical people, but at some point, you'll find yourself coming up with analogies to explain a complex technical concept to someone who isn't as technical as you are.

We deal with a lot of abstraction in our day-to-day jobs and deal with things like entities, instances, classes, interfaces, modules, and so many other concepts that are hard to articulate and explain to other people—sometimes even other programmers. Even though these concepts may make sense in your own head, finding the right words to verbalize your thoughts is sometimes difficult.

The most important thing to keep in mind in these conversations is to respect your audience, especially if they aren't able to follow along when talking about technical concepts. The last thing you want to do is to be condescending because they don't understand a complex technical topic that seems like second nature to you.

Try your best to avoid using slang, acronyms, or technical jargon. Instead, try to use metaphors or draw comparisons to common concepts that almost any audience would understand. Trying to connect a technical concept to common everyday objects or ideas that your audience may be able to relate to can even be fun and challenging.

In some situations, it may be beneficial to teach your coworkers simple programming concepts if you know you'll be working closely on a project in the future. Teaching them concepts such as an API or the difference between frontend and backend systems can go a long way in helping them understand requirements or limitations for a project. Additionally, they may end up teaching you about marketing, operations, customer success, or other topics related to the team they work on. The more knowledge you have about different organizations within your business, the better decisions you'll be able to make when designing systems and user interfaces and solving problems for your customers.

While you may be interacting with people from different areas of expertise, you may also be interacting with people of varying seniority and experience, so it's important to keep your audience in mind so you communicate at the appropriate level for your audience.

Another way to gauge the audience's communication needs is to ask them directly. Sometimes directors, vice presidents, or executives may not care about the details of how you implemented your solution, but sometimes they actually are interested in how it works under the hood. The best way to determine how detailed to get is to straight up ask them how detailed you should get when explaining something. Most likely all they care about is that the problem was solved or the bug was fixed, but sometimes they might want to know the nitty-gritty details about it.

Additionally, they may be more interested in knowing *when* something will be done or if something is even possible to do and less interested in *how* it's done. The role of upper management is to set the vision, put together a timeline, and prioritize resources to bring the vision to life. Oftentimes, they may not care what design patterns or frameworks you're using, but they do care about how long a project will take so that they can fit it in with the other priorities on the roadmap.

The bottom line is that programmers with good communication skills are the ones with the ability to communicate complex ideas in clear and concise ways to both technical *and* nontechnical audiences. In doing so, they can build rapport with their team members, including individual contributors, managers, directors, VPs, and executives. When your team members know they can trust you to convey important information to various audiences, you will be seen as a valuable asset to the company, which could lead to more responsibility and a higher salary.

12.3 *Writing Skills*

Once you've determined who your audience is and how you should approach the conversation, the next thing to be aware of is the channel you're using to communicate. Whether the conversation takes place in person or through a written form of communication such as chat or email will determine how you approach the conversation.

The first channel we'll dive into is written communication. As a programmer, you'll be reading and commenting on a lot of code reviews throughout your career, and how you communicate your thoughts and ideas in writing can determine how well those ideas are received by your coworkers.

CODE REVIEWS

Code reviews are often one of the poorest areas of communication between programmers, which is unfortunate because it's also one of the most important. Constructive comments from senior developers are one of the best tools to help you become a better programmer because you're receiving direct feedback on how you can improve your code and your logic. Whether it's more concise syntax, bringing clarity to your logic, ensuring maintainability, improving performance, or handling edge cases you didn't think about, the act of having someone else peer review your work will improve your technical skills.

Unfortunately, code reviews can often be intimidating experiences for developers because it's easy to feel like the reviewer is criticizing the programmer and not their code. The wrong words or the wrong tone can do more harm than you may realize, so it's important to be cautious of this when reviewing and commenting on another's code. It's always best to stick to constructive feedback, and you should never personally attack the programmer for coding a solution a certain way, even if it's wrong.

There are numerous ways to solve programming problems, and the best solution may not always be apparent to the person assigned to the task. While it may seem easy to disagree with someone's code because you believe there's a better way to solve the problem, criticizing their thought process or their implementation is the wrong way to go about it.

To build healthy relationships with your coworkers, it's imperative that you stick to constructive criticism during code reviews. It doesn't help anyone if your comment that the code is wrong or that it's inefficient. Instead, explain *why* their solution is suboptimal and, more importantly, offer advice on *how* they can improve their code.

Let's look at examples of two different types of comments you may see during code reviews, and why one is better than the other.

This comment doesn't help the programmer at all because it doesn't explain why the code will cause a performance issue, and it doesn't offer a solution for how to fix the issue. It's possible the programmer was lazy when writing their algorithm, but it's also possible they may not even be aware their code is inefficient. The commenter comes off as abrasive, and the programmer who wrote the code may take their criticism personally, which could lead to indecisiveness and self-consciousness the next time they submit their code for review.

First, it's less aggressive if you approach the comment as a question, rather than a statement. That way, you're asking the programmer if they had considered your concerns, and it gives them an easy way to admit they hadn't thought about that aspect when coding their solution, without trying to embarrass them.

Next, the comment offers a suggestion for how the programmer can improve their solution. Perhaps they were so caught up in getting to a working solution that they completely missed the opportunity to improve the efficiency of their algorithm. That doesn't necessarily mean they're a bad programmer by any means, so don't make them feel like they are.

Another aspect of this comment is that it avoids using the word "you" and replaces it with "we" and "us." By doing so, you're signaling to the programmer that they are part of a team and that you're all in this together.

Senior engineers tend to focus more on the success of the team as a whole, rather than just themselves, and it's the small details like this that go a long way if you want to show that you're a team player. Try not to single out programmers when offering feedback on their code. You're all on the same team, and it's everyone's responsibility to deliver the highest quality code possible. It's always better to stress the importance of the team and acknowledge that the code is owned by everyone and not a single person. In doing so, you'll build confidence in your teammates and they'll take your feedback seriously.

Now let's look at the next way in which you'll communicate with others through writing, which is in your project management system.

⚙ RESOURCES

- How to Deliver Constructive Feedback in Difficult Situations[124] (productivityhub.org)
- How to Review Code as a Junior Developer[125] (medium.com/pinterest-engineering)

---

124. https://productivityhub.org/2019/04/19/
    how-to-deliver-constructive-feedback-in-difficult-situations/
125. https://medium.com/pinterest-engineering/
    how-to-review-code-as-a-junior-developer-10ffb7846958

12.3.2 TICKETING SYSTEMS

There's a good chance your team is using some kind of bug tracking or project management system. These tools help teams plan, organize, and track new features, tech debt, and bugs as tickets make their way through the development process and into production. Ticketing systems can get complicated very quickly with multiple people moving tickets around, creating subtasks, setting deadlines, and marking certain tickets as high priority.

Many programmers seem to despise these tools because they are heavy on process and always seem to get in the way. There are complex workflows, planning meetings, and coming up with arbitrary estimations for tickets—all of which sometimes feels unnecessary. While it may seem like these tasks take time away from actually writing the code they're meant to track, these tools are important for project managers to stay organized and even more important for the management team to track progress on larger projects and to plan out longer-term goals that align with the company's strategy.

Given the complexity of these tools, it's important to make sure you're communicating clearly on your tickets as you move them through the pipeline. When you create a new ticket, provide as much detail as possible. Don't assume that other developers on your team will understand where to find the bug, how to reproduce it, or why a certain part of the code should be considered technical debt and needs to be cleaned up.

Software engineers can be lazy now and then; we've all been there before. There's nothing worse than pulling in a ticket to work on and finding almost no information to go on. There may be a title and a one-sentence description. How do you know how to reproduce the issue if it's a bug? How do you know if it's a high-priority feature request from an important customer? How do you know if it's tech debt that's not really an issue but the developer who filed it felt like the code was just a bit messy? It's sometimes impossible to know these things when there's no information to work off of, and as a developer, it's frustrating when you are assigned these kinds of tickets.

There's almost no information on this ticket! There's no context to work off of when it comes time to plan the next development cycle, except that a bug exists. It's not even clear if the bug still exists because there's no instructions on how to reproduce it.

Furthermore, it's possible this particular report is complex, with multiple export settings, and the bug only exists for a specific configuration setting. If you pick up the ticket and try to reproduce this bug, the report's export functionality might work fine for you because you don't have the application in the same state as the person who reported the issue.

There's also no indication of any acceptance criteria, which is a set of predetermined requirements that must be met in order for the product owner to mark the ticket as complete. In other words, how will we know the bug is actually fixed? It's important to lay out this information up front, before the development work begins, so there is a clear finish line to know when you're done. Doing this helps reduce scope creep and keeps the developer focused on solving the original problem instead of making additional unrelated changes in the process.

So, how would you improve this ticket so the issue is clearly communicated?

You can already see that there's much more information in this ticket that will help prioritize this work during planning sessions and will help the developer reproduce the bug when they pick up the ticket. The ticket provides a stack trace to help the developer pinpoint exactly where the error is thrown on the server, as well as steps to reproduce the bug. Notice how the description states that the bug happens when exporting in CSV format, which is also helpful. Without that, someone may try to reproduce the bug and be able to successfully download the report in Excel format.

Lastly, the acceptance criteria are clearly defined so the project manager knows exactly what requirements need to be met in order to mark the story as complete. This helps the developer focus on fixing the actual bug rather than optimizing the code or cleaning up a part of the codebase unrelated to the actual bug itself.

You'll be creating lots of tickets throughout your programming career, so it's important to build good habits now. While they may seem trivial to you at first, these small details make a big difference. Junior developers tend to overlook this aspect of software development because they're focused on writing code. Sure, writing code is important, but keeping information organized in the issue tracking system is another piece to the puzzle when it comes to documenting, planning, and delivering software.

When writing tickets, it's helpful to think of it as if you're communicating with someone in the future. Whether you're communicating with yourself or someone else, your goal should be for whoever picks up the ticket tomorrow, next week, or next year to have all the information they need to work on the task. If they pull up your ticket a month from now and there's no information except for a title, there's a good chance they'll just close the ticket and move on to something else because they don't know enough context to pull it into the next development cycle.

Which brings us to the next point—closing tickets. While your version control system acts as a historical record of changes made to your codebase, your project tracking system acts as a historical record of all implemented features, bug fixes, and technical debt that has been paid down over the years. It acts as a paper trail for the work your team has accomplished and also offers insights on how efficient your team was in delivering the work.

Leaving notes and details about each ticket (and updating any relevant documentation) as the work happens allows project managers and new engineers to look back and get important context for why some action was taken on a ticket. A new engineer may find a reference to a ticket number in a git commit or a comment in the codebase that they can pull up and get additional context around why a specific piece of code was written.

It's important to add comments and give good reasons when closing tickets, because other people may find that information useful in the future. Sometimes multiple developers will report separate tickets for the same bug, or there may be tickets that overlap in scope. Other times the change may be so low in priority that it's closed to reduce the clutter during a ticket grooming session. Other times the ticket may not be relevant anymore because the bug has already been fixed or priorities have changed. Regardless of why the ticket is no longer a priority, it's always a good practice to add a comment giving a reason for why you're closing the ticket and choosing not to work on it.

Here are a few examples:

⟩ EXAMPLE

- "Closing this ticket because it is a duplicate of BUG-394."
- "Closing this because the issue is no longer reproducible. It looks like it may have been fixed as part of the refactoring done in ENG-1138."
- "Closing this since it will no longer be needed once we migrate to the new ordersservice."

The worst thing you can do is to close a ticket without any context because whoever pulls up that ticket in the future will have no idea why this work was never done. It takes less than a minute to add a one-sentence comment about why you're closing a ticket, so try to document why you're doing something when making changes.

And lastly, not everyone will be able to monitor the comments and activity for each individual ticket, so it's helpful to tag the relevant people when adding your comments. It's easier to keep track of ticket activities when you get email or chat notifications when things happen, rather than having to continuously monitor each ticket for new comments or movements in their status.

Following these few tips while working in your project management system may not seem like much, but your manager will appreciate good communication skills, even in asynchronous channels. In general, the more information you can provide, the better, especially when it comes to updating others on the progress of specific tasks in your tracking system. As you work your way towards a senior role, building good habits when working in your project tracking system will let everyone benefit from the knowledge and information you share.

12.3.3 COMMUNICATING OVER CHAT

Chat systems have been around for decades and have been adopted by companies of all sizes, from startups to public companies. They've even become valuable tools for open-source projects with the rise of Slack, Microsoft Teams, Discord, and many others. While you're probably already experienced with chat communications, here are a few things to keep in mind when using them a professional setting. It may be simple and obvious advice, but it goes a long way when exchanging information with others.

Asynchronous

Similar to tickets in your project tracking system, communicating over chat systems is asynchronous in nature. Although chat is much more real time, you may or may not get a response right away. Conversations can happen quickly in group chats, or they can happen over the span of hours or days. If you need an immediate response, consider other channels of communication. If you're in the office, talking to someone in person is the quickest way to get the answers you need. If you're working remote, try to get them on a video call or at the very least, a voice call. Recognizing which conversations are best held over chat and which ones are better off in a different channel will help avoid miscommunication and will get you the information you need in a timely manner.

Acknowledge requests

One habit that is good to get into is to simply acknowledge when someone asks you to take a look at something. You may need to ask someone to take a look at your pull request or to review a design document, and it's sometimes frustrating if you need an answer right away but your request just sits there without any response.

When someone asks you to review something, give a quick response like "Taking a look", "Looking now", or "Will take a look in 15 minutes after this call" to let them know that you saw their request when you will look at their document. When someone reaches out to have you look at something, it's usually urgent enough to warrant asking for a review. This small habit lets the other party know that you're working on it so they aren't met with complete silence on their end.

Choose the proper channel

Different channels are meant for conversations around different topics. Your company may have a few different channels like #general, #marketing, #engineering, #industry-news, or a #random channel. The larger the company, the larger the audience in each channel, which is why it's important to respect the channel topic when posting or holding a conversation with someone. It's insensitive to hold personal conversations with others in public channels. If possible, try to move those kinds of conversations to a direct message channel with the person you're talking to.

On the other hand, sometimes it's best to hold certain conversations in public rather than in a direct message. If it's an urgent conversation or a discussion about a topic that multiple people should be aware of, then it would be helpful to discuss in a public channel, because it allows others to follow the conversation along with any decisions that are made.

12.3.4 COMMUNICATING OVER EMAIL

By now, we're all comfortable communicating over email, so I won't get too deep into details here. Similar to communicating over chat, try to determine if email is the right channel for the kind of information you're trying to communicate. Before sending an email, consider other communication channels and determine if it's better to ping someone over chat or hop on a video call. Sometimes a 15-minute video call is more efficient and effective than multiple emails back and forth with other people.

Urgency is also another thing to consider and may dictate which communication channel is best. Sometimes it's best to pick up the phone or hop on a video call to discuss something time-sensitive. Alternatively, if the conversation isn't time-sensitive, email could be a good asynchronous option.

Regardless, email is still widely used throughout the world, and there are certain conversations where it's the best option, such as communicating with people outside of your organization. Unfortunately, email conversations tend to be slow, and important points can be easily misinterpreted. Let's look at a couple of things you can do to improve your communication over email.

12.3.4 Subject Lines Matter

Use the subject line to get your point across. Lots of people scan subject lines when they're busy or in a hurry, so try to make yours specific and to the point so that someone scanning their inbox will know they need to open your email.

Always try to be as specific as possible, while keeping the subject line concise. A good subject line should hook the recipient in and stand out among the clutter of an inbox.

Better subject lines are clear, to the point, and worth the extra few seconds it takes to think of something specific. They go a long way in helping make sure your emails are read and not skipped over by someone skimming an inbox.

12.3.4 Keep It Short

Some people get hundreds of emails a day, which leads to email fatigue. When things are moving quickly in the office or when there's a time crunch to meet a deadline, not everyone will have the time to read every email they get. People frequently scan emails to look for relevant information, and a good way to guarantee your email doesn't get read is to send a big wall of text to your recipients.

Keep emails short. The shorter the better. And try to get your point across in the first or second sentence. This helps people that are in a rush and gets them the relevant information they need. Not everyone will read to the end of your emails, so if you're asking someone a question in the third paragraph, don't assume they will see it and respond with an answer.

12.3.4 Use Formatting

Additionally, a good way to make your emails easier to scan and make your most important information stand out is to use text formatting for your important points.

- Use bullet points or numbered lists to make it easy to convey different options or lists.
- Use bold fonts to stress importance. The bold lettering will usually be the first thing someone reads, so use it sparingly and only for the most important one or two points you're trying to make.
- If you have to write a long explanation, break it up into multiple paragraphs. There's nothing worse than a huge wall of text, and multiple paragraphs makes it easier to read.
- Consider underlining or using font colors to highlight the most important information. If you do this, do it consistently within a single email. Keep in mind that certain colors can be hard to read if they don't contrast well with the background. And using colors to highlight some text can draw attention away from other parts of the text which may be overlooked.

  - Use red to highlight disclaimers or risks that are critical for people to know. You can even preface the sentence with "Disclaimer:" or "Important:" in boldface or red to add more impact.
  - If you're responding to an email and answering someone's questions, highlight your answers in a different color like blue. This makes it easier to differentiate between the original author's questions and your answers.

12.3.4 Use @ Tags

In group emails, if you need to direct a question to a specific person or direct specific people to handle action items, it's helpful to use @ tags. It's usually a good idea to combine these with bold formatting to make sure your requests are seen.

- **@dave** do you still have that infrastructure diagram from last quarter?
- **@allison** can you coordinate with marketing on a release date?

When someone sees you've tagged their name, you're more likely to drive action or get a response from them, but as always, use them sparingly otherwise they lose their effectiveness.

12.3.4 Don't Send Immediately

Whatever you do, *do not* send the email immediately after you finish writing. There's no undo once you hit send, so taking a few minutes to proofread your email could save you from embarrassment in the future.

Sending emails with grammatical errors can actually hurt how people perceive you and how much they trust your opinions, so it's crucial to catch any mistakes before hitting send. There are tools, such as Grammarly,[126] that can highlight grammar mistakes in real time and help you catch simple mistakes and clarify your writing.

Even if you're using tools to help catch errors, you should still proofread your own writing. Reread your email, and then reread it again. Try to combine or remove sentences completely in order to shorten the email to just the necessary information. Only after you've proofread your email multiple times should you hit send.

Now that we've covered some things you can do to improve your written communication, let's look at verbal communication.

12.4 *Speaking Skills*

Speaking in front of a group of people can be very uncomfortable, especially for programmers. In fact, public speaking is one of the most common and stress-inducing fears there is, regardless of which profession you're in. Almost all people experience anxiety before they have to speak to an audience, so you're not alone if your nerves get the best of you.

While you may never need to speak in front of your entire company, you may find yourself in one-on-one conversations, team meetings, or larger all-hands meetings where you're asked to speak on a certain topic.

---

126. https://www.grammarly.com/

Improving your public speaking skills has many benefits and will help you be more effective at your job.

Let's take a look at some things you can do to improve your speaking skills, whether you're talking to one person, a small group of people, or a large audience.

12.4.1 NERVES ARE NORMAL

Speaking publicly doesn't come naturally to most people. No one wants to embarrass themselves in front of an audience, yet that's the most common fear among people who have to get up and speak in front of others.

It takes a lot of practice and preparation to build confidence in your public speaking skills, but even the best public speakers still get hit with anxiety and need to manage their nerves. In fact, some even argue that nerves can be a good thing because the adrenaline rush makes you more alert and helps you focus on what you need to communicate. Regardless of how you deal with the anxiety of speaking in front of people, learning to manage your nerves is a valuable skill to build. You may never be able to get rid of your nerves, so learning to manage them will help you communicate more effectively.

12.4.2 PREPARATION IS KEY

If you get up in front of others to give a talk without any kind of preparation, you've failed to set yourself up for success. Finding time to prepare for the talk will make a huge difference in your confidence, and you'll have more time to figure out how to explain your topic clearly and concisely.

Research your topic. You want to make sure you have a thorough understanding of what you'll be speaking about. To teach others about a topic or idea, it's crucial to have a solid understanding of it yourself. If you don't understand something well, you'll have a hard time teaching others about it.

If you're putting together a presentation with slides, add notes to each slide. These are usually talking points you want to touch on for that slide. Add bullet points and use short sentences or phrases so it's easy to scan during your presentation.

If possible, ask someone else to proofread your slides, especially if you will be speaking to people outside of your immediate team. Spelling mistakes, grammatical errors, and inconsistencies in your talking points can distract listeners from the message you're trying to get across, so it's good to catch these things early with a second pair of eyes, similar to how you have your coworkers proofread your pull requests when making code changes.

Practice running through your slides at least once before giving your talk. Find an empty conference room in the office or do it at home in front of a mirror. Talking out loud helps you find parts of your presentation that may sound awkward or don't make sense. It'll help you identify which points you're having trouble explaining so you can refine your notes and improve your delivery.

Sometimes, you'll only have a limited amount of time to give your talk. It could be five minutes or fifteen minutes, but you want to make sure you respect the time of others who may be speaking after you. If you practice beforehand, make sure to time yourself so you have an idea of your talk's length. This will help you know if you need to shorten it and cut things out of your presentation, or if you're under time and can dive deeper into one or two key slides.

12.4.3 STAY ON TOPIC

Staying on topic sounds easy, but it can be difficult when you're in the middle of a speech or presentation. It can be intimidating standing in front of a group of people who are staring back at you, and you'll feel like you need to say something interesting to keep their attention. Try not to ramble on and talk about things that are not closely related to your topic. Try to get your point across as succinctly as possible, which is why practicing beforehand is crucial to refining your message.

More words do not necessarily mean more information, so don't assume that the more you talk the more your audience will learn. In fact, rambling on will often dilute your message and make it harder for your audience to understand the points you're trying to make. Most people will tune out after a certain point or once they reach a certain information threshold, so it's usually better to keep it as short and sweet as possible.

12.4.4 ANTICIPATE QUESTIONS

Always try to leave a minute or two at the end to answer any questions people may have. Depending on the complexity of the topic and how familiar people are with it, you may have quite a few questions from the audience. It's usually a good idea to anticipate what questions people may have, but it's impossible to think of every question you may be asked.

Also keep in mind that you may not have the answer to all the questions people will ask, and that's okay! It's not a bad thing to admit that you don't know the answer, but let them know that you'll follow up and try to get them an answer later on after the presentation. Then, make sure you actually do follow up and try to answer their question as best you can.

So, now that we've gone over specific things to keep in mind when communicating both in writing and verbally, let's look at some more general things that will help you become a better communicator.

⧉ RESOURCES

- 10 Tips for Improving Your Public Speaking Skills[127] (harvard.edu)
- Important Public Speaking Skills for Workplace Success[128] (thebalancemoney.com)

12.5 *Listening Skills*

Most people assume that communicating is all about how you write and speak to other people, but that's only half the story. Communication also requires you to listen to other people's thoughts, ideas, and concerns, and that's equally as important for collaborating and working well with others.

Just as it can be frustrating when you feel like you're not being heard, your teammates, coworkers, clients, and customers will also feel frustration if you don't listen to what *they* have to say. Building rapport with others requires mutual respect between all parties, and everyone's thoughts should be taken into consideration with equal weight. Other people's opinions matter just as much as yours do, so make sure to listen to what they have to say.

So, what can you do to become a better listener?

12.5.1 REFLECTIVE LISTENING

One of the easiest things you can do is to practice reflective listening. The whole idea of reflective listening is quite simple. First, listen to the speaker and try to understand their ideas, then try to paraphrase the idea back

---

127. https://professional.dce.harvard.edu/blog/
     10-tips-for-improving-your-public-speaking-skills/
128. https://www.thebalancemoney.com/public-speaking-skills-with-examples-2059697

to them. You don't have to repeat what they said word for word, but try your best to summarize their ideas in order to confirm that you understood them correctly.

While it may feel silly at first, there are a few benefits to reflective listening:

- It shows the speaker that you're listening. This goes a long way in building mutual trust with the speaker.
- It helps to point out any misunderstandings, because they will be apparent in your summary that you repeat back. It will also help the speaker formulate and clarify their ideas.
- It will help you embrace the speaker's perspective without forcing you to necessarily agree with it. It helps open up your own opinions to new ideas without having to commit to them.
- It's especially helpful when discussing business processes and procedures from other organizations within the company. For example, if you're working on a project to integrate two systems, it will be easier if you have a holistic view of how they work. Ask the speaker to explain the processes you're tasked with automating, and repeat it back to them to make sure you understand it correctly.
- It will help you clarify assumptions during the requirements gathering phase when planning new features and projects.

Reflective listening is a technique you can put to use today that will automatically improve your communication skills, and chances are you may already be doing it to some extent. As always though, be careful of using this technique too much or going too far with it, as it can come across fake or forced.

🔗 **RESOURCES**

- Reflective Listening[129] (wikipedia.org)
- How To Practice Reflective Listening (With Tips And Examples)[130] (indeed.com)

---

129. https://en.wikipedia.org/wiki/Reflective_listening
130. https://www.indeed.com/career-advice/career-development/reflective-listening

12.6 *Prepare for Meetings*

Not all meetings are created equal. While your daily stand-up meeting may not seem too important, you will be in other meetings where crucial decisions are made. Depending on the topic of the meeting, it may be worth your time to prepare so you have an idea of what you want to communicate before you need to. That may involve scanning your project management board to remember what you worked on yesterday, reading through the codebase to refresh your memory about how a particular component works, or reviewing documentation for potential third-party services.

You may be called upon in the meeting to give your opinion or your input on how a particular part of the system works and how easily it can be extended. If it's not fresh in your mind, it may be hard to give a complete answer during the meeting when everyone is relying on your input. By preparing ahead of time, you will be able to give an answer confidently so that important decisions can be made.

It's also good to go into meetings with a list of predetermined questions you'd like to have answered. Don't assume that everyone is on the same page about how easy or hard some change will be, or how the change should be made. Oftentimes, other people haven't considered a solution you may be asking about, so asking the question can be helpful to others as well as yourself.

Finally, it's good to keep a notebook and write down your thoughts and questions before starting the meeting. Conversations happen quickly in meetings, and if you try to keep it all in your head, you may not remember everything you wanted to bring up. Writing down notes also helps organize your thoughts, and you can cross things off your list as they are discussed.

These may be simple ideas, but as they become second nature, these habits will help you communicate your thoughts more clearly throughout the course of your career.

12.7 *Don't Communicate through Other People*

Did you ever play that game called "Telephone" growing up? It's a game where kids stand in a circle and one player whispers a sentence to the person next to them. The second player then repeats the message to the third player, and so on. When the message reaches the end, the last player announces the sentence that was whispered to them and compares it to the original sentence from the first player. Almost always, the two sentences are completely different due to each player interpreting and repeating the message with slight differences to the next person. With each iteration the message becomes less like the original.

The same thing can happen in professional settings as well, even if it's unintended. If possible, try not to rely on someone else to pass your message along to the intended recipient, because it may not be communicated exactly as you intended it to be. Sometimes though, you may not have an option, such as if you need to convey important information up the management chain to the executive team. In general, the more people your message passes through, the higher the chance it will be misinterpreted by the receiver.

If possible, send a chat message, an email, or speak to the recipient directly rather than communicating through a chain of people. If you must pass along information through others, try to follow up with the recipient and confirm they got your message. It may seem trivial, but it's yet another habit you can build now that could save you from headaches in the future when collaborating across teams and organizations.

12.8 *Communication Isn't Just "Saying Something"*

Good communication is about being able to convey your ideas in ways that are properly received by your audience. It's simply not enough to assume that just because you said something people will understand your ideas. You may not always be able to get your point across, which can be frustrating as a programmer when it comes to conveying technical ideas to your teammates.

Additionally, just because you tell someone something doesn't mean it's no longer your responsibility. As programmers, we're prone to the bystander effect when it comes to the maintenance and operations of our systems. Individuals are less likely to take responsibility for something when there are other people present, because they assume someone else will step up to the plate and take care of what needs to be done.

A better approach is to not only communicate what the problem is, but also communicate what actions should be taken, and if possible, come to an agreement on *who* will take responsibility.

The best communicators are successful because they have the ability to inspire actions that lead to certain outcomes. Senior engineers tend to learn this skill as they work to develop their teammates and improve the codebase. They don't just assume that everyone will understand what needs to be done or how something should be done, and they know that just raising concerns about something doesn't mean it will get done. If you want something to get done, you'll need to learn how to inspire action.

12.9 *Dealing with Conflicts*

If you stick with programming long enough, you'll eventually be a part of some emotional discussions. As programmers, we take pride in our craft, and it can be easy for individuals to get attached to certain solutions or architectural designs. You'll deal with conflicting views at some point in your career, and emotions may run high.

It's okay to disagree with your teammates, but how you handle yourself will speak volumes about your character and how your teammates view you. In fact, healthy debates are a sign of a high-functioning team, but the discussions must be respectful. While it's good to debate the pros and cons of different designs and algorithms, it can be bad if things turn from a civil conversation to a full-blown argument.

In rare cases, passionate developers may get into tense arguments over which solution is the better approach. It's possible that each approach has its strengths, weaknesses, and trade-offs, and that both developers are correct. No solution is perfect, and that's okay.

If you find yourself in a heated discussion and you sense that emotions are running high, the best thing you can do is to keep the conversation as civil as possible. Take the higher road if possible, which sometimes means making compromises. It may even be best to table the conversation and walk away to let everyone cool off. You can always pick up the conversation again at a later time once people have the opportunity to reflect and think things over some more.

Always remember that it may take months or years of hard work to build up trust between your coworkers, but you can lose that trust in an instant if you lose your temper during a heated discussion. You won't always be able to convince people that your ideas are the best, even if you feel like they are, and that's okay. A good quality of a senior developer is that they realize that no solution is perfect and they sometimes have to make decisions on suboptimal solutions.

Pick your battles, because you don't want to lose the trust of your teammates over the name of a variable or which design pattern to use to solve a problem. Being a senior developer means making compromises, and sometimes it's best to just go with the flow every now and then.

12.10 *Wrapping Up*

While it may be difficult to realize when you're just starting out, poor communication skills often contribute to programmers plateauing in their career. As a programmer in a senior position, you will lead technical projects and mentor younger developers. To continue on the trajectory to staff and principal engineering roles, you'll need to learn how to build support for your ideas and work cross-functionally with nontechnical people across your company. And if you choose to go down the managerial path, good communication skills are even more critical, as you'll be managing projects and people constantly.

Complex software systems cannot be built by one person alone. Modern-day software solutions require multiple people, both technical and nontechnical, to collaborate and deliver products that meet evolving customer needs. Successful teams consist of a broad set of people with diverse backgrounds and skill sets, and the ability to connect, collaborate, and solve real problems with different people is a rare skill that is often overlooked by programmers.

Once you reach a certain technical level, everyone will have the necessary skills to solve the problems at hand in some way, but not everyone will have the communication skills to convey their ideas and gather feedback when they need to. The bottom line is that the higher up you advance in your career, the more you will stand out if you are an excellent communicator. The best programmers communicate with empathy and listen to what others have to say, and those who communicate the best will be the first to advance when it comes time for a promotion.

13 Work-Life Balance

As programmers, we spend a lot of time in front of a computer. It's not uncommon to go a full day staring at pixels on a screen as you click and type away. There's a lot of pressure from employers to work long hours to reach the quarterly and annual goals set out by the management team, and it always feels like there's too much work and not enough resources. The deadlines are tight, but we have to ship this quarter!

There are competing interests between employees and employers that may be hard to understand early in your career. When you're young, you're just happy to have a job and a good salary. But as you grow more experienced and progress through the different stages of life, your priorities may change.

Your employer, whether you like it or not, is motivated to run a streamlined and efficient business. Unfortunately, your employer's goals probably don't align with your long-term goals. Your company is incentivized to squeeze every ounce of productivity out of you while paying you as little as possible. Businesses operate on margins that they are naturally incentivized to increase by keeping costs low.

On the other hand, as a young programmer and individual contributor, your incentives are a bit different. Your goal is to maximize your salary, thereby increasing your quality of life, all while working as little as possible. Who wouldn't want to make more money while working less?

So, now you can see there are conflicting interests between a business and their employees, and whether you like it or not, the business usually holds most of the power when it comes to the negotiating table. They are able to push their employees to work harder and meet tight deadlines because there are only a few options for the employee:

- Work hard for a promotion.
- Meet expectations enough to stay employed.
- Fall short of expectations and get fired.
- Leave voluntarily to find a better opportunity.

Most people choose to work hard for a promotion, but sometimes that promotion involves a lot more work than you may realize. It's possible you may not love the work you're doing but are happy with your comfortable salary, so you show up every day and do what's asked of you, but nothing more.

Because you're reading this book and you've made it this far, you probably fall into the camp of ambitious programmers who are looking to work hard for a promotion, which is great. The early years of your career are exciting and filled with possibilities and different directions for you to choose. You're like a sponge—ready to absorb as much knowledge and experience as you can. You'll spend long hours tracking down a pesky bug or building out the UI for a new page because it's fun. And it won't always feel like work because you love to write code and build things and solve problems.

But there are a lot of negative side effects from working too hard.

There are other things outside of work that are at least equally as important as working long hours to meet your deadlines. In this section, we'll explore what a good work-life balance means, and what you can do today to ensure you'll have a long and healthy programming career.

13.1 *Time Is Your Most Valuable Resource*

You sell your time to your employer in exchange for money in the form of a salary. That might sound weird at first, but this is true for most people who work for a living. For people earning an hourly wage, it's pretty straightforward: The number of hours they work directly correlates to the amount of money they make. More hours equal more money.

> ❯ EXAMPLE

Let's look at some numbers:
Hourly wage: $50 per hour
Earnings per week: 40 hours x $50 per hour = $2,000 per week before taxes
Earnings per year: 52 weeks x 40 hours x $50 per hour = $104,000 per year before taxes

Pretty simple—more hours worked equal more money earned. For people on a salary, however, it's a bit different. You and your employer agreed to a predetermined salary, usually on an annual basis, when you signed your employment agreement. If you're on a salary, you're not paid directly for the number of hours you work, but you're expected to put in a certain amount of work and produce a certain amount of output.

> ❭ EXAMPLE

Let's assume a 40-hour workweek:

Earnings per year: $104,000 before taxes

Earnings per week: $104,000 / 52 weeks = $2,000 per week before taxes

Hourly wage: $104,000 / 52 weeks / 40 hours = $50 per hour

On a $104K annual salary and assuming a 40-hour workweek, an hourly employee and a salaried employee will earn the same amount. But unlike someone earning an hourly wage, if you're on a salary and work longer workweeks in order to meet deadlines, the numbers aren't in your favor.

> ❭ EXAMPLE

Let's look at what happens if you work a 50-hour week on the same salary:

Earnings per year: $104,000

Earnings per week: $104,000 / 52 weeks = $2,000 per week

Hourly wage: $104,000 / 52 weeks / 50 hours = $40 per hour

And what about working a 60-hour week?

Hourly wage: $104,000 / 52 weeks / 60 hours = $33.33 per hour

While there are many benefits that come with a salary, the reality is that your income for the year is fixed, which means you'll be paid the same regardless of how many hours you work. Unlike an hourly wage, more hours do not equal more money. In fact, more hours equal less money on a *per-hour* basis. It's not because you're earning less, but because you're using more of your time to earn the same amount.

⬦ CAUTION Just to be clear, I'm not advocating for slacking off or working as little as possible to maximize your earnings per hour. My goal is simply to illustrate the relationship between salary and hours worked per week so that you are more conscious about how they affect each other. Knowing this will hopefully encourage you to work smarter when you are in the office so that you can avoid long nights and weekends when you need to hit a deadline.

In some cases, however, you may have no choice but to put in extra hours. As programmers, we deal with issues when our software fails, and it can fail at any time. You may get a phone call or a chat notification that the server is down, and you'll need to hop on the computer in the middle of the night to help get the server back online. Or you may be working on a big project and need to have a working demo before an investor meeting, so you may have some long nights during crunch time. Lastly, you may need to work long nights and weekends so you don't get fired. If you can't get your work done during normal business hours, your job may be at risk.

There are a number of good reasons why you *should* work more than 40 hours a week, and everyone's situation is different, so it's something you'll need to figure out on your own. It's okay to work long hours every now and then, but when you find yourself working long hours week after week, it will start to affect your work-life balance.

It's important to be aware of when your work starts to affect your personal life. If you're canceling plans with friends or family, haven't taken any time off in months, or haven't been able to find any time for your hobbies because you're too busy with work, try to take a step back and reflect on your work-life balance.

◇ IMPORTANT Your job does not define your life.

We program for a living, but that doesn't mean we should be coding all day every day to earn that living. It's important to create a life outside of work for your own mental health, and to build relationships you can lean on if needed.

You weren't meant to stare at a screen your whole life. There are plenty of benefits to getting away from the computer and unwinding in the analog world, and in the end, you'll want to look back on your life and the incredible memories you made, not that you wrote the perfect algorithm or solved a tough programming problem for your employer.

It's good to focus on increasing your salary early in your career, but you'll soon realize that with a higher salary comes greater responsibility. You'll be responsible for keeping projects on schedule, keeping systems up and running, and keeping your team's throughput high so that they can continue to ship code to production, and so much more. A lot of these responsibilities don't involve any coding, and some of them will involve sitting in more meetings and spending more time planning and writing feature specs and bug reports. You may have long nights and weekends rotating as the on-call engineer, or triaging and fixing bugs in the middle of the night.

The responsibilities that come with a higher salary are not the most glorious parts of being a programmer, and not everyone is cut out for them. You may not like doing these things, but you'll justify it because the pay is so good. What's important to realize though is that higher pay does not always bring happiness. What's the point of sitting in meeting after meeting if you're not happy at the end of the day? Is it worth the higher salary if you're missing out on experiences with friends because you have too much work?

At some point, you'll need to decide what's best for yourself and your mental health. You'll need to find a balance between your work life and your personal life, and that may mean taking a lower salary if it means you'll be happier and have more time to spend with friends and family. Or it may mean that you stick to the individual contributor route rather than the manager path so that you can continue to write code. It's up to you to figure out what makes you happy.

You'll need to decide if it's really worth it to spend late nights in the office, or if you'd be happier with less responsibility and the freedom to head home earlier in the day. Again, it's okay to work late hours occasionally if it means keeping an important project on schedule, but if you find yourself working late week after week, even when there are no upcoming project deadlines, you may want to step back and ask yourself if that is really what you want.

13.2 *More Hours != More Work*

A common misconception is that the more hours you spend in the office, or working from home late at night, the more work you'll get done. While it may feel this way, it can actually have the opposite effect and lead to a negative impact on the *quality* of your work.

Working longer has diminishing returns, because at some point your brain will hit a wall where you'll start to drift and lose your ability to focus. A 60-hour workweek is not the same as two 30-hour workweeks. While it may feel like you're getting more done in half the time, it may be lower-quality work.

Instead, focus on working smarter and more efficiently during your workweek so you can get all your work done in 40 hours. Distractions and context switching can kill your productivity.

⟩ EXAMPLE

Here are some examples of what you can do to stay focused:

- Buy a nice pair of noise-canceling headphones to minimize distractions.
- Block off chunks of time on your calendar for "focus time."
- If possible, try to stack meetings back-to-back rather than having them spread throughout the day so you're not switching contexts constantly.
- Block social media and news websites on your work computer so you're not tempted to check them during the day.

The more efficiently you can work during normal business hours, the less you'll need to work after hours.

🔗 RESOURCES

- Preventing   burnout   for   programmers[131] (medium.com/@karolisram)
- How to Prevent or Recover from Developer Burnout[132] (dev.to/ actitime)
- 83% of Developers Suffer From Burnout, Haystack Analytics Study Finds[133] (usehaystack.io)

13.3 *The Work Will Always Be There*

Sometimes, you'll find yourself working on a problem towards the end of the day and it'll feel like you're making good progress. You'll want to keep the momentum going and will feel like working a little late to wrap things up. You've almost got your code working; just a few more lines of code and everything should compile without errors.

At some point, you need to find a good stopping point and just call it a day. For your own sake, it's better to close up your laptop and unplug at the end of the workday. You can always pick up where you left off tomorrow, or

---

131. https://medium.com/@karolisram/preventing-burnout-for-programmers-12b4968adbaa

132. https://dev.to/actitime/how-to-prevent-or-recover-from-developer-burnout-3g5f

133. https://www.usehaystack.io/blog/
      83-of-developers-suffer-from-burnout-haystack-analytics-study-finds

next week, so don't put too much pressure on finishing a task before heading home for the night. The work will always be there tomorrow.

It's a marathon, not a sprint, so you need to set stopping points and take breaks from the computer. Good software takes years to build, so you're never going to get it all done in one day or one week. Rome wasn't built in a day, as they say.

We're never done building software. There's always something that can be improved, whether it's fixing bugs for more reliability, implementing a faster algorithm or a better user experience, or reducing the cost of our infrastructure. There will always be more work to be done, so don't put so much pressure on yourself to stay late and finish what you're working on.

You won't be able to hit every deadline, and it's not the end of the world if a project runs over schedule occasionally. Software projects are notoriously hard to estimate correctly, and sometimes your estimates will be wrong. In the end, quality software that takes a little longer is better than buggy software that was rushed through to production, so keep that in mind when you're working late.

An important aspect to a good work-life balance is to keep a regular schedule. It's easier to close up your computer at the same time every day than it is to decide at different times when you're done working for the day. You may notice some engineers pack up their computer once the clock hits a specific time because they know how important it is to spend time away from their work. They know the importance of stepping away from the computer and of having a meaningful life outside of the office.

It's equally important to find a routine in the morning that works for you. Try not to rush into work as soon as you roll out of bed, which can lead to stressful and hectic mornings. It's important to find time in the morning to relax before getting your workday started. Maybe that involves doing yoga, reading the news with a cup of coffee, spending time with your family or your pets, working on a side hustle, or exercising. Spending time to relax before the stress of your workday is important for your mental well-being and a good habit to build if you want to perform well.

13.4 *Side Projects*

Writing software for work can be fun. You get to use cool technologies, and you get paid to solve tough technical problems. But writing software for fun can also be satisfying. When you're at work, you don't get to make every single technical decision, but when you work on side projects, you have a blank canvas. You can build whatever you want, however you want. It's refreshing, satisfying, and frustrating all at the same time. Experiment and try new things, and don't worry if the code gets messy because you don't have other engineers peer reviewing your work. You're able to cut corners in order to get something to work quickly, and this is where your creativity really shines, because there's no risk of failing.

The excitement you get when you take an idea in your head, build it with code, and see it come to life is hard to describe. Lots of developers enjoy it so much that they'll work on their own projects outside of work. It's easy to get immersed in these projects because coding often doesn't feel like work. We do it because it's fun and we love the challenge of problem solving.

But it can be hard sometimes. Not because you can't solve a problem, but because it often feels like there's a stigma in our industry if you're not working on a side project or contributing to an open-source project, especially when you're applying for new jobs. There's pressure to work on side projects so that you have some work to show a potential new employer, especially if you're just getting started in your career and don't have a lot of professional experience.

It's completely okay if you don't have any side projects or any open-source code you feel proud of to show off. You shouldn't feel any pressure to contribute to the open-source community. If you do, that's great, but don't feel like you're any less of a developer because you don't code in public.

Contributing to open-source software can often be intimidating for inexperienced engineers. You may not feel like you understand the code enough to contribute, or you may be embarrassed to put your code out in the public—and that's okay. The open-source community can be pretty harsh, and people often have unreasonable expectations for how software should work, even when it's free.

Some of the best programmers in the world have never written a single line of code for open-source software. There are a number of reasons why this may be.

- They are forbidden by their employer. Sometimes, the contract you sign says you cannot share any code, algorithms, or learnings from your day job with anyone else.
- They don't have time to write open-source software because they'd rather spend their free time with their friends, family, and pets.
- They don't know what to build.
- They have other priorities in life outside of programming.

Whatever the reason may be, don't feel like you absolutely need to work on side projects or write open-source software to be a good developer. It's completely optional and should be something you choose to do because it brings you joy or maybe because you're lucky enough to make some side income from a project.

In the end, working on side projects outside of work is a double-edged sword. It's easy for some people to get home from work, open up the laptop and get right back to coding. Sure, it's a good way to learn new technologies and level up your skills, but don't let coding take over your life. You're a programmer by trade, but don't let that define who you are. It's good to step away from the keyboard every once in a while, and honestly, you should find time every day to do things that don't involve staring at a screen.

⚙ RESOURCES

- Do Software Developers Normally Code on Weekends?[134] (codeahoy.com)
- Developers' side projects[135] (joelonsoftware.com)

13.5 *Friends and Family*

We're all human, and we all need meaningful interaction with other human beings. It's right there, sandwiched in the middle of Maslow's hierarchy of needs. After our physiological and safety needs have been met, humans have a need for interpersonal relationships and a feeling of belongingness. This social belonging that we all strive for can come from friends, family, or a significant other.

---

134. https://codeahoy.com/2019/10/19/do-software-developers-work-weekends-work-life-tech/
135. https://www.joelonsoftware.com/2016/12/09/developers-side-projects/

As you learn to build a work-life balance that works for you, it will be important that you find time to break from the digital world and work on forming new friendships and relationships. And it's equally important that you nurture your existing relationships with friends and family. We're social creatures, and it's good for our mental health to unwind with people we care about.

If you find yourself in a new city or feel like you're lacking friends at any point, it's up to you to make an effort to change that. Try to get out of your comfort zone a bit and meet new people. It's extremely hard at first, but the more you do it, the more you'll feel comfortable with it.

> ❯ EXAMPLE

Here are some ideas for meeting new people:

- If you're close with your coworkers, ask them if they'd like to hang out after work or on the weekend. You already know them well from working together, so try to find any shared hobbies or activities that you'll enjoy together. Some people personally don't like to mix work with pleasure, but other people have met lifelong friends and partners who started out as coworkers.
- Attend local meetup groups in your city. If meeting other people is hard for you, try attending some technical meetups first to get your feet wet. You'll be around like-minded people, and you may even learn about a new technology, language, or framework while you're at it. It's even better if you can find and attend some nontechnical meetups related to a favorite hobby or interest. You'll meet some interesting people, and as you attend more meetups, you'll start to recognize familiar faces and get to know people better.
- Join a sports league. Adult sports leagues are popular, and your employer may even sponsor some. You get the advantage of getting exercise in addition to meeting other people and getting out of the house.
- Volunteering can be rewarding in more ways than one. Not only are you giving back to your community, but you'll be working with other people who want to help too.

These are just a few examples of what you can do in your free time to get out and meet other people, but it doesn't have to always be in person.

Even a phone call to your parents, siblings, or close friends can give you a mental boost and lift your mood after a long day of work. It's easy for us to let our relationships dwindle as we get older, but a little bit of effort to keep in touch with people close to you can go a long way.

13.6 *Take Care of Your Body*

Programming is mentally taxing, but it also has a big impact on our physical health. You may be able to get by on pizza, energy drinks, and sleepless nights when you're young, but as your career develops, you'll need to put more focus on taking care of yourself in order to stay sharp. Your health is crucial to your well-being, and issues with your health may contribute to issues in your career and personal life.

If you take care of your diet and your physical health, improvements in your physical and mental health will follow. A few changes in your lifestyle can snowball into daily habits and routines that can pay dividends well into your future, and it doesn't necessarily take a big time commitment each week to see results.

> ≫ EXAMPLE

Improving your diet and physical fitness will:

- Increase energy levels
- Improve mental health
- Help with weight management
- Strengthen your bones and muscles
- Reduce chronic pain
- Reduce the risk of depression
- Reduce the risk of disease
- Help improve your sleep

The hardest part about improving your physical health is finding motivation, especially if you have a busy schedule and are under a lot of stress at work. Building up to a consistent routine takes time and dedication, but once you get the ball rolling, it will be easier to stay motivated and continue.

≫ EXAMPLE

Here are some ways to find motivation:

- Look for opportunities to make small changes to your lifestyle. For example, go for a walk after each meal to help your body digest instead of sitting down at the desk after lunch or on the couch after dinner.
- Set aside time in your day for physical activities. Block off time on your calendar to go for a walk in the morning or a run during lunch. Set it up on a recurring schedule to help build a daily or weekly routine.
- Start with activities and locations that you enjoy. Take a hike on your favorite trail with great views, or bike around your favorite neighborhood. Making sure you enjoy what you're doing and where you're doing it will help you stay motivated.
- Try exercising with friends. They can hold you accountable if you try to skip a workout, and you can hold them accountable as well.
- Use fitness trackers as a way to gamify the process and compete against yourself or your friends.
- Start slow, and only work up to more intense workouts when you feel like you're ready. It's okay to progress at your own pace, and there's no need to compare yourself to others who may be farther along in their fitness journey.

Exercise and physical activities are an excellent way to improve your mood, your health, and your overall well-being. It doesn't matter what kind of exercise you choose, whether it's cardio or weight lifting. What's more important is that you get up and move. Your body will thank you for it, and it'll help keep you feeling young.

⚭ RESOURCES

- Overcoming Barriers to Physical Activity[136] (cdc.gov)

---

136. https://www.cdc.gov/physicalactivity/basics/adding-pa/barriers.html

13.7 *Reading*

Reading is a great way to unwind at the end of the day, and it's an easy way to avoid staring at a screen before going to bed. There's nothing better than getting lost in the pages of a book that you can't put down.

It doesn't matter what kind of books you read; the only thing that matters is that you're reading. In doing so, you open your mind up to new ideas, new characters, new feelings, and new ways of thinking. You can get lost in a fictional world with vivid scenery and a protagonist, or learn about the history of a famous figure that changed the world. You can improve your confidence with a self-help book, or even expand your skill set with a good technical book. Just pick a book or a topic that sounds interesting to you and start reading.

The hardest part is actually finding the time to read. For busy professionals, good times to read are in the morning right when you wake up, or at night right before going to bed. Try to find time in your schedule that works for you. The good thing about books is that you can take a break and pick up right where you left off, so even if your schedule is busy, you can still read when you're able to find some time.

The more you read, the more you'll come to realize what you enjoy and who your favorite authors are. Writing well is a difficult skill, and when you find an author you enjoy reading, they can take you on journeys you never thought possible.

13.8 *Traveling*

We all need time away from work to unwind and relax. There's no better feeling than closing up your computer on Friday afternoon after a long week. The weekend has arrived, and you've got a few days off to do whatever you want. Time to get out of the house and explore your city.

It's exciting to find a new restaurant you like, a new park with a great view, or a hidden gem within your city. What's even better, though, is traveling to a new city where every restaurant, park, or public space is waiting to be discovered.

Traveling to new destinations can be fun, relaxing, and inspiring, and sometimes stressful, but don't let that stop you. Everyone should travel because it opens you up to new experiences and challenges you to get out of your comfort zone. You'll need to plan ahead, make decisions on the fly, stay organized, and get to places on time in order to make your connections. Not all of these skills translate well to your professional career, but they give you good experience and help build confidence that you can solve problems on your own.

Traveling for an extended period of time helps to clear your mind so you can come back to the office feeling refreshed and ready to jump back into your work. It helps break up the monotony of your job when the weeks start to blur together. An upcoming trip will give you something to look forward to, and you'll have memories and stories to share when you get back and are ready to jump back into work.

If possible, try to travel internationally, even if it's just to a neighboring country. It'll open up your world to new and exciting perspectives. If you've never traveled abroad before, it can be intimidating, but afterwards, you'll be glad you did. You'll experience new cultures and societies and see how they solve daily problems similar to ones you experience at home. You'll learn about their history and experience their art and architecture.

Don't be intimidated if there's a language barrier; it'll force you to get creative with how you communicate as you try to convey your thoughts. Sometimes, it can be surprisingly easy to understand the gist of a conversation, even if it's between two people that don't speak the same language. Using hand gestures and simple words and phrases can get you most of the way there. These new skills will help you throughout your career and give you a better sense of empathy when communicating with other people.

It's a very humbling experience when you return home from traveling abroad. You come back with new perspectives on the world and a new respect for different cultures. You'll have more appreciation for the things you have and a better understanding of things you might be taking for granted. Traveling can be rewarding and one of the best things you can do to grow as a person. You'll learn things that you can use both in your career and in your personal life, and you'll come back with some great stories to tell.

🔗 RESOURCES

- 5 Scientifically Proven Health Benefits of Traveling Abroad[137] (nbcnews.com)
- 15 Benefits of Traveling and Why Travel is Good for You[138] (cabinzero.com)

13.9 *Wrapping Up*

It's entirely up to you to create the work-life balance that's best for you. No one else can help you do that, so you'll need to figure out what kind of balance you really want. If your current employer's values and policies don't align with the work-life balance that you're looking for, it may be time for you to look for one whose does.

If there's one piece of advice you take away from this section, you should always remember that you are a programmer by trade, but that should not define who you are. It's not healthy to spend all of your time in front of a computer screen typing into a text editor. That's not the way we were meant to spend our lives, so get out of the house and away from the keyboard as much as you can.

You won't regret it one bit when you're older and look back at the memories you made along the way. There's more to life than just work, so have fun and make it a life worth living.

🔗 RESOURCES

- Work/life balance will make you a better software engineer[139] (codewithoutrules.com)
- Life as a developer : Balancing work and life[140] (codeburst.io)
- Facts and myths about work-life balance in software development[141] (softwaremind.com)

---

137. https://www.nbcnews.com/better/wellness/
     5-scientifically-proven-health-benefits-traveling-abroad-n759631
138. https://www.cabinzero.com/blogs/our-journey/benefits-of-traveling
139. https://codewithoutrules.com/2016/11/10/work-life-balance-software-engineer/
140. https://codeburst.io/life-as-a-developer-balancing-work-and-life-787d463393fa
141. https://softwaremind.com/facts-myths-work-life-balance-software-development/

14 Asking for the Promotion

We've covered quite a bit by now, and it's okay if it feels overwhelming. You're not expected to know all of these things right now, and it's impossible for someone, even a senior developer, to be an expert on *every* topic covered in this book.

If you take what you've learned throughout this book and put it into practice each week, you'll come to a point where you'll be confident in your technical abilities and ready to ask for the promotion to a senior role. Asking for a promotion can be daunting, but if you feel like you're ready to make the jump, it's important to start having the conversation with your manager.

It may sound easy, but it's harder than a lot of developers realize. You'll be putting yourself out there and asking your manager to evaluate your technical abilities and your soft skills. If you're feeling vulnerable, that's normal. After all, you're putting your career trajectory in the hands of someone else. Of course it's nerve-racking!

In most cases, it will be up to you to get the conversation started. Yes, your manager *should* be thinking about the career development of their developers, but that's not always the case. You shouldn't assume that the business will automatically reward you for doing your job well. Sometimes your hard work may be recognized, but there's no guarantee anything will happen automatically. Sometimes it takes a little nudging to get the ball rolling.

To an extent, you will always need to sell yourself in order to get what you want. It's best to assume that if you don't sell yourself, no one else will. It's up to you to promote your accomplishments and to demonstrate to the management team that you are able to deliver results consistently and that you're worthy of a promotion. Don't just assume your track record speaks for yourself and that people will notice.

So, let's look into some things to keep in mind when you're gearing up to ask the big question.

14.1 *Preparing*

First off, it's important to do your homework. You'll want to get familiar with your company's process. Every company handles promotions differently, and the process may differ depending on the maturity and size of your company.

Informal promotions. In smaller companies, it's common for the managers to have direct authority over deciding who gets promoted. The managers may meet with each other and present the candidates they feel are most deserving before coming to an agreement on who should be promoted.

Semi-formal promotions. In growth-stage startups and midsize enterprises, a semi-formal process is common. The leadership team will begin to add some structure in an effort to standardize and encourage fairness in the decision-making process. The managers may still meet together to discuss potential promotions, but the candidates are evaluated against certain criteria, rather than individual managers' opinions.

Formal promotions. More formality is common among public companies and large established enterprises, and in some cases, it can be a very long process. Candidates are evaluated against well-defined criteria for each level of the job ladder. You'll most likely be asked to put together a self-review and potentially gather reviews and recommendations from your peers. Your manager will also write their own review of your performance. There will often be a committee of senior engineers and engineering managers that will evaluate your reviews and recommend a promotion if they feel you meet the bar.

It's important to understand what kind of process your company uses to promote employees so that you can prepare correctly. You'll want to make sure you have your ducks in a row before starting the process, and knowing how you'll be evaluated is crucial.

14.2 *Assess Yourself*

It's common for companies to promote employees who are already performing at the level they're being promoted to. The actual promotion is a recognition that your work output is exceeding expectations at your current level, and it's a nod from the leadership team that you're ready to consistently deliver results at the next level.

So, in order to assess yourself, you need to determine how you're doing in regard to what your manager expects from a senior engineer. This is much easier if you're in a company that has a clearly defined leveling framework. It's a good idea to read through the expectations for your current level first. Take notes on each criterion in your current level and give yourself a grade for how well you think you're meeting that expectation.

🔥 CONFUSION In almost all cases, leadership will expect you to perform very well at your current level in order to even be considered for a promotion at all. If you're not meeting expectations at your current level, it's going to be very difficult to meet them at the next level. Additionally, in some places you'll be expected to already be performing at the promotion level, not just at your current level.

It's important to be honest with yourself during this step. Writing down that you're exceeding expectations for every criterion at your current level doesn't help you at all. No one is perfect, and there is always room for improvement. You'll be disappointed if you get your hopes up only to hear from your manager that there are still some areas that you need to work on.

Next, do the same thing for the next level up, which should be a senior role. You may feel like you are already meeting some of the expectations for the next level, which is great! But don't get discouraged if there are areas at that level you still need to work on. If you're honest with yourself about where you're not meeting expectations for a given criterion, you will at least have an idea of what you need to work on, and you'll be able to put together a plan with your manager to build those skills.

14.3 *What to Prepare*

Even if you're not required to put together a formal self-review of your performance, it's still a good idea to do it anyway. You'll be able to use this to build a compelling case for why you think you deserve a promotion, and it will give you some good talking points when discussing the promotion with your manager.

Back in the You're not an Impostor[§4] section, we talked about keeping a log of all your previous accomplishments in a notebook or notes app. This is the perfect time to pull up that document and review everything you've accomplished in the last 12 months, because that will give you plenty of ideas.

What should you write down?

- Examples of successful projects. Think of the highest-impact projects you've worked on and stress the importance of the role you played in making the projects successful.
- Concrete examples about how you improved some metric or key result, and the impact that it had. Give actual numbers for the metrics if possible.
- Examples of how you demonstrated leadership and showed the maturity that is expected from a senior role. Try to find examples of where you put the team first and took action to lift up the entire team.

Your goal with this document is to prove that you are already working at the level that you're asking to be promoted to. The more examples, the better, but make sure you choose them selectively and purposefully. Think about it as a resume for your last 6–12 months at your current role.

14.4 *Build Relationships*

Building relationships in the workplace is always important, but especially when it comes to asking for a promotion. First and foremost, you'll want to get your manager on your side, because it'll be nearly impossible to get a promotion without the support of your manager.

How do you do that?

- Do your job well. If you're not meeting expectations for your current level, it'll be hard to convince your boss that you deserve a higher salary and more responsibility.
- Ask your boss to explain the promotion process at your company. This will let them know that you're interested in advancing your career, but there's no pressure to actually ask for the promotion yet.
- Ask them what their philosophy is on promoting their employees. This will give you an idea of what qualities your manager is specifically looking for in order to recommend a promotion.

- Ask them to rate your performance against your current level, and against the next level up. This is a way to have an informal conversation around what criteria you are currently meeting, and what criteria you still need to work on.

These conversations should happen during your one-on-ones where you can talk privately and openly with your manager. You should approach the conversations as a way to gather information and feedback without actually asking the big question yet.

Additionally, it's good to talk to other senior engineers on your team or on different teams. Even though they are your peers, senior engineers often have at least some sort of say in the promotion process. Your manager may lean on the senior engineers to give their honest feedback about your performance when it comes time to build a case for your promotion.

If you have good working relationships with the senior engineers, they'll be more likely to put in a good recommendation for you when they're asked for feedback. Plus, you can ask the senior engineers if they think you're ready for more responsibility. If you have a good relationship, they may feel more comfortable talking candidly with you about what you're doing well and what you need to work on. The senior engineers have a unique perspective because they work more closely with you than the managers do. Your manager may be busy throughout the day, and they won't be able to read every line of code that you write. A senior engineer on the other hand may have a better understanding of certain technical or soft skills you may need to work on before asking for a promotion, and your manager may even consult the senior people on your team about recommendations for who to promote.

14.5 *Asking the Question*

Once you've put in the work to prepare and you feel like you're ready, it's time to ask for the promotion. To be clear, asking for a promotion is not a one-time conversation. Rather, it's an ongoing conversation between you and your manager about what you need to do to be considered for the promotion. Very rarely will you be promoted on the spot, so it's going to take some time and hard work to get across the finish line. Don't expect things to happen overnight. Asking the question is just the first step to get the ball rolling on the process. So, how should you ask?

One way to ask for a promotion:
"I'd like to be considered for a promotion to a senior software engineer. I feel like I've demonstrated that I'm ready based on my recent performance, but I know there are still some areas that I can improve. I'd like to start an ongoing conversation with you to identify what I need to work on in order for me to reach the next level. What do I need to do to show you that I'm ready for the next step?"

The goal is to ask for the promotion, but not demand it. Humbleness goes a long way here, so it's good to acknowledge that you know there are still some areas where you can improve. The idea is to work with your boss to put together a roadmap for what you need to work on to get to a point where they're confident in your abilities and comfortable going to bat for you when submitting their recommendation for your promotion.

Keep in mind that your manager is evaluated on their decision-making skills and their team's performance, so there's a strong incentive for them to recommend a promotion only when they believe you are ready. If your manager recommends a promotion to their boss when you're not ready to perform at the next level, it may make them look bad and hurt their credibility.

There will almost certainly be some areas that your manager would like to see you improve before they will be comfortable enough to recommend a promotion for you. Work with them to develop a plan to improve those areas. They may be able to put you on projects that will give you certain experience, or give you more responsibility that will help you develop skills they want to see you improve.

14.6 *Put in the Work*

Once you've got the ball rolling and have a plan in place with your manager, it's up to you to put in the work to show your boss that you're serious about the promotion. Set weekly goals for yourself to work on the areas that your manager would like to see you improve. Write them down and review your progress regularly, like at the start and end of each week.

Some things your manager expects of you won't be easy. You'll be forced out of your comfort zone, and you'll be asked to do things you've never done before. You may need to make important decisions, whether they're technical ones or choices about how to handle certain processes.

You may not always make the right decisions, but the important thing is to learn how to think logically, creatively, and collaboratively with the rest of your team. Being a senior engineer is about taking on more responsibility and putting the team first, and your boss may set you up to gain experience in making bigger decisions before you actually get promoted. If you can demonstrate that you're able to make decisions and lead within your team, you'll be able to show your manager that you're ready for the promotion.

If you take what you've learned throughout this book and put it into practice each week, asking for the promotion shouldn't be that intimidating. The hard part is actually applying what you read here to the real world. Not everything is as black and white as it's made out to be, but the learning curve is where you grow from a junior engineer to a senior engineer.

If you don't get a promotion the first time you ask for it, it's not the end of the world. Get as much clarity on why. It may be other factors in the company. If it's that they think you're not ready, ask them what areas you need to focus on improving in order to be considered. You're on a journey of continuous improvement, and you'll never stop learning about how to improve your craft.

Just because you meet the requirements for a senior role doesn't mean you'll automatically get a promotion. But you're much less likely to get something you never ask for—and you are missing an opportunity to grow and learn.

🔗 RESOURCES

- How To Ask for a Promotion[142] (hbr.org)
- How To Ask for a Promotion[143] (jocelyngoldfein.com)
- Software Developer Promotions: Advice to Get to That Next Level[144] (blog.pragmaticengineer.com)

---

142. https://hbr.org/2018/01/how-to-ask-for-a-promotion
143. https://jocelyngoldfein.com/how-to-ask-for-a-promotion-87e0e3b4ebd6
144. https://blog.pragmaticengineer.com/software-engineering-promotions/

- How Software Engineers Get Promoted — Career Advice[145] (becomebetterprogrammer.com)

15 Conclusion

By now, you should have a much deeper understanding of what it takes to be a senior software engineer, and you have a roadmap for what areas to work on to prepare for your next review cycle. If you've learned something new from this book, that's great! That means I've done my job. My hope is that this won't be the last time you pick it up. I'd encourage you to revisit any section in the future if you need a little refresher or you'd like to refer back to any tools or techniques you learned along the way.

In Growing Your Career,^{§2} you learned about the different career paths available to you, whether you choose to pursue the individual contributor path or the management path, and which one may be the best path for you. And don't worry, you have plenty of time left in your career to make that decision, so there's no rush right now.

We covered what differentiates a senior programmer from a junior one in What Makes You a Senior Engineer,^{§3} along with what areas you should focus on developing in order to demonstrate that you're ready to perform at the next level. These qualities take time to build and some trial and error to get good at it, but in the end it's worth it.

In You're Not an Impostor,^{§4} we discussed impostor feelings. You learned to remember that everyone deals with impostor feelings, even the senior- and staff-level engineers, and management too. Hopefully, you picked up some techniques for how to handle those times when you're feeling insufficient, and how to use it as a learning opportunity to fill in your knowledge gaps.

In Working with Your Manager,^{§5} you learned how to approach working with your manager so you can build a strong working relationship with them. When you have a foundation built on trust, you'll be able to be open and honest when you need to be, and you'll learn how to adapt your working style to their management style. Having a good foundation with your manager helps during high-stress situations and will go a long way when it comes time for your review cycle.

---

145. https://www.becomebetterprogrammer.com/how-software-engineers-get-promoted/

Next, in How to Recover from Mistakes,[§6] you learned that all developers make mistakes, and more importantly, you learned some tools and techniques that you can use to identify and resolve incidents quickly. You should now have the knowledge to handle critical incidents in a professional manner and to work with your teammates to prevent the same mistakes from happening twice.

Not all questions are created equal, and in How to Ask Better Questions,[§7] we talked about how asking good questions is a skill that takes effort on your end before asking your teammates to take time out of their day to help you.

You learned how to read unfamiliar code[§8] and quickly get up to speed on an unfamiliar codebase—something you'll need to do many more times throughout your career. Being able to jump into a new project and provide value quickly will help build trust with your teammates and confidence in your own technical abilities.

We talked about how to add value,[§9] not only for your own organization but for the customer as well. Consider this your North Star for your career—if you can find ways to build value, you'll be rewarded with a successful career.

In How to Manage Risk,[§10] you learned the different ways that risk can creep into the software development lifecycle, along with tools and techniques to manage and mitigate those risks. Hopefully, by now you have a much better understanding of what to watch out for and what things can contribute to added risk. You also learned that some risk is acceptable, and that it's impossible to completely eliminate all risk, so it's better to learn to live with it instead.

You learned tools and techniques for delivering better results,[§11] and what you can do to help keep yourself moving forward when you run into roadblocks and dead ends when building programs. By now, you should know when to consider trade-offs when it comes to writing software, and that no codebase will ever be perfect.

In How to Communicate More Effectively,[§12] you learned about what considerations you should take into account when communicating with your team and across teams. Depending on the type of communication and the urgency of the conversation, you should be able to choose the best channel and communicate your ideas clearly and concisely. You also now know that listening is equally important as speaking when it comes to effective communication.

While you may love what you do, your job is not your life. In the section on work-life balance,[§13] we discussed that to be good at your job you need to first take care of yourself, both mentally and physically. Don't forget to foster friendships, both new and old, and take some time off to travel, relax, and learn some new hobbies.

And finally, you learned how to prepare for a promotional cycle and how to approach a conversation with your manager to ask for a promotion[§14] to a senior role. Don't wait for the performance review cycle to focus on building the skills you need. The time to start working on those skills is now. If you're able to build habits and work towards continuous and incremental improvement, you'll be qualified for a promotion in no time.

About the Author

David Glassanos is a seasoned software developer with over a decade of experience building successful consumer and enterprise products for international brands. David started his career in Silicon Valley at Mertado, an e-commerce startup acquired by Groupon in 2012. While at Groupon, he built systems to personalize emails delivered to 100 million inboxes each day. From there he joined NatureBox as an early employee, and helped grow revenues to $50 million annually. Since then he has helped build products and engineering teams in the e-signature, co-working, and transportation telematics industries. As a way to give back to the developer communities that have helped him succeed, he advises junior software developers on career growth, effective communication, soft-skills, and work-life balance.

About Holloway

Holloway publishes books online, offering titles from experts on topics ranging from tools and technology to teamwork and entrepreneurship. All titles are built for a satisfying reading experience on the web as well as in print. The Holloway Reader helps readers find what they need in search results, and permits authors and editors to make ongoing improvements.

Holloway seeks to publish more exceptional authors. We believe that a new company with modern tools can make publishing a better experience for authors and help them reach their audience. If you're a writer with a manuscript or idea, please get in touch at hello@holloway.com.

www.ingramcontent.com/pod-product-compliance
Lightning Source LLC
Chambersburg PA
CBHW030506210326
41597CB00013B/808